THE FUTURE OF CAPITALISM

ROBOTS, ARTIFICIAL INTELLIGENCE AND EMPLOYEE STOCK OWNERSHIP PLANS

ROBERT C. HACKNEY

License Agreement
Limits of Liability and Disclaimer

Published by Four Palms Publishing

ABOUT THE AUTHOR

Robert C. Hackney is a lawyer, author, and entrepreneur. He lives in Jupiter, Florida. As an entrepreneur, Mr. Hackney started and sold an internet company in the early 1990s, formed a small telecommunications company, owns part of a diverse healthcare company and, with a partner, owns a legal software company. He is the author of numerous books, including *Mergers & Acquisitions 101; Entrepreneur's Guide to the New Equity Crowdfunding Rules; Let's Bring America Back; and Lawyer's Guide to Blockchain Technology*, among others.

He attended the University of Miami, received a Bachelor of Arts degree from Florida State University, and a Juris Doctorate degree from Florida's oldest law school, Stetson University College of Law. Upon graduation from law school, he went to work for a state agency that regulates a number of businesses, primarily in the securities industry. Leaving government service, he joined a large law firm headed by a former United States Senator. His practice is now divided among his two offices in Jupiter and Stuart, Florida.

His legal career over the past four decades has been broad. He has held every position from serving as a managing partner of a large law firm with four offices, to working as a sole practitioner. He has handled matters ranging from multimillion dollar mergers to murder trials. His broad corporate law experience has ranged from representing "Mom and Pop" type companies to major domestic and international corporations. He has been involved in mergers involving Fortune 500 companies as well as small private companies.

Mr. Hackney is a member of The Florida Bar, the United States District Court, Southern District of Florida, the United States District Court, Middle District of Florida, the United States Court of Appeals for the Fifth Circuit, and the United States Court of Appeals for the Eleventh Circuit.

He has lectured and authored publications related to business and legal topics. Mr. Hackney has been active in many civic and community organizations, and has served as President of the Marine Industry Education Foundation, President of the Rotary Club of Palm Beach Gardens, Florida, President of the Guild of Catholic Lawyers of the Diocese of Palm Beach, founder of the Alternative Energy Association, and is listed in the Who's Who Registry.

Dedication

This book is dedicated to my wife, Cyndee Kay.
She knows why.

TABLE OF CONTENTS

PROLOGUE

We are at a crossroad. The next two decades will decide our long term future in America, and the world. Our leaders, and in particular our press, needs to stop getting bogged down in senseless bickering about subjects that have no real benefit to any of us. We face issues that are not being discussed and problems that need to be overcome, for the benefit of America's economic future.

How will we create new jobs and new industries? How will we fix our infrastructure problems? Will technology help or hurt us over the next two decades? What is our long term plan? Do we even have a short term plan?

How will education change due to technology? What kind of robot is going to take my job? What kind of new jobs will be invented? With the explosion of online education available, will any of it help with training young people for the future?

With family income decreasing as it relates to cost of healthcare, housing, college and other basics, how will we stem the tide of the increase in income inequality? Will the gap continue to drive the middle class lower and lower? Does anyone really believe in the concept of universal basic income?

CHAPTER I

ARE THE ROBOTS COMING TO TAKE MY JOB?

Let's not kid ourselves here, robots already run most of our world. We'll be their butlers soon enough.

Eric Stoltz

There are two obvious answers to that question. One is that it is a "definite maybe," and the other is that it depends upon your job. The reality is, however, yes, robots are coming and they may either help you in your job or they may replace you. Either way, they are coming, and some are already here.

According to a report issued by the Office of the President of the United States in December 2016, robots have made the economy more efficient. The President's Report cites an analysis of twelve developed economies, which includes the United States, that concluded that artificial intelligence could double annual economic growth rates by 2035 in the countries studied.

Economic pundits refer to three industrial revolutions, ie: driven first by steam power, then electricity and finally electronics. The World Economic Forum has referred to a fourth industrial revolution, which is says will be driven by artificial intelligence.

As of this writing, however, artificial intelligence based automation has not yet had a major impact upon growth in economic productivity. I expect that to change, and such change may be rapid.

Thus far we have mixed the terms "robots" and "artificial intelligence." We will separate and distinguish them as we go forward. Robots are generally the physical and mechanical devices that are driven by artificial intelligence, or AI. I like to think of it as the AI being the brains, and the robot being the body carrying out the orders of the brains.

As we all know, technological advancement is nothing new in America. In the mid-1800s, about one-half of all Americans were involved in the agricultural segment of the economy in one way or another. Today that number is about 2% of the working population. As such, we have adapted, changed and evolved with the times. Now America produces more than enough food for domestic consumption, and has adequate resources to export massive amounts of food to other countries, all with only 2% of our population involved in agriculture.

As beneficial as the technological advances in agriculture were to the overall economy, these advances came at the expense of many individuals who were displaced and had their lives and jobs disrupted in the process.

Predicting the changes that will come with more and more AI automation has become a hobby for many industry analysts. The reality, of course, is that all predictions are inherently flawed, as we will discuss in a later chapter.

We can, however, feel comfortable in our view that there is certainly a trend and that the trend will grow. Where it will impact us the most we cannot determine, but there are obvious places that seem to be already attracting attention. One focus has been on self-driving vehicles. One example of this technology is Google's self-driving car. Google started its work on this project in 2009, and by 2016 had logged over one billion simulated miles and over 2 million actual miles driven. Google states that it had accumulated a combined 300 years of driver experience by the end of 2016. Google's parent, Alphabet, Inc., created a company called "Waymo" in December of 2016 to continue on with the self-driving car project with the mission of making such vehicles available to the public.

Google is certainly not alone, Elon Musk's Tesla, Uber, Toyota, Mercedes Benz and GM are all working on their own versions of autonomous vehicles.

While having an individual vehicle for your own use that drives itself is compelling, the greater implication for the economy comes in the use of commercial vehicles like taxis and trucks.

In 2016, Uber introduced its first tests of self-driving taxis, which are running around Pittsburgh, San Francisco, Phoenix and other cities. According to the US Bureau of Labor Statistics, there are approximately 240,000 taxi cab drivers in America. Consequently, there are potentially almost a quarter of a million jobs on the line if self-driving taxis become a reality (and they will). Uber itself has seriously impacted the taxi business, and in its next phase it will impact those individuals who have become Uber drivers.

In the commercial area of trucking, America has between 1.7 and 3.5 million truck drivers that will eventually be impacted by self-driving trucks.

Have you met Otto? He is quite a driver, and he isn't even human. Otto is a company based in San Francisco which has developed AI technology that it can retrofit to trucks so that they can drive themselves. Otto is still in the testing stages and so is not fully developed, but at the pace of technological change, it won't be long before there will be Otto driven trucks on the highways of America. At the moment, Otto's "brains" cost about $100,000, but the company expects this price to reduce to the area of $10,000 per unit.

In October of 2016, Otto carried 2,000 cases of Budweiser beer down Interstate 25 from Fort Collins, Colorado to Colorado Springs, Colorado, or about 100 miles. A backup driver was in the truck, and it was preceded and followed by Otto personnel.

This was the first real demonstration of the potential of the technology and also showed its present limitations. To get on and off the highway, the human driver had to handle the driving, since Otto cannot yet drive in city traffic.

Otto's founders are two former Google employees, Anthony Levandowski, formerly of Google's self-driving-car project, and Lior Ron, who was the head of Google Maps. Other former Google employees joined them and the company was created in 2016.

In late 2016, Uber acquired Otto. The marriage of Uber and Otto gave Otto's team the ability to work with about 500 Uber engineers who are devoted to self-driving technology. With Otto, Uber now can focus on transportation of people and delivery of goods.

Of course, Otto is not alone. It is reported that Volvo Trucks, Peterbilt and Daimler Trucks are all in the process of developing systems for self-driving trucks.

If you drive a taxi, or if you drive a truck commercially, I have two words for you: "drone school."

I know, it sounds funny, but the skills you have as a taxi driver or a truck driver may translate to flying drones. According to a 2013 report by the Association for Unmanned Vehicle Systems International, there could be more than 100,000 new jobs created by 2025, and those jobs would have an economic impact of $82 billion. The Association for Unmanned Vehicle Systems International is a group representing 7,500 individual members and 600 corporations.

If you are looking for a new career, there will be opportunities for people with drone training that include military drone pilot, firefighter, disaster relief, search and rescue, law enforcement, oil and gas operations, seismic study, border patrol, traffic reporting, storm chasing, agriculture, package delivery, forestry, engineering, computer science, commercial contractors, film, and other industries.

To meet this need, many colleges and universities have created programs, including Embry-Riddle Aeronautical University in Daytona Beach, Florida,

Oklahoma State University in Stillwater, Oklahoma, Liberty University in Lynchburg, Virginia, University of Nevada, Reno in Reno, Nevada, Monroe Community College, Corporate College in Rochester, New York, and Mohawk Valley Community College, Center for Corporate and Community Education in Utica, New York, among many others.

Interestingly enough, the Unmanned Vehicle University in Phoenix, Arizona was created solely for the purpose of drone training. One of this school's programs is the drone pilot training course that totals 46 course hours, broken down into 3 sections. The first 16 hours cover the basics of drone topics, the next 10 hours will be spent flying over 40 different drones in a simulator, the final 16 hours will be real flight training at a specified center. The school also has doctoral and master's degrees focused on the engineering and program/project management of drones.

The drone market was $11.3 billion in 2013 and is projected to be $140 billion in 2020.

The Story of Mary

Mary is 87 years old and lives in a nursing home. She is hearing impaired and nearly blind. She is no longer able to read, or even watch television. Mary is somewhat feeble, and is in a wheelchair. In addition to her other troubles, she suffers from dementia. While she recognizes family members, and can clearly talk about things that happened fifty years ago, she has a very bad short term memory. During some of her days, she gets into a "loop" where she asks the same questions over and over again.

What is this place? Do I live here? How long has I been here? Mary is somewhat typical of many people her age. Due to her situation, her biggest problems are boredom and depression.

Recently, things have gotten much better for Mary. She has a new friend named Pepper. Pepper is a great friend because she is attentive, kindly and never gets bored or angry with repetitive questions from Mary. Pepper has learned Mary's personality traits and is responsive to them. Pepper has adapted to Mary so much that she knows all of Mary's tastes and habits. She knows what Mary like to eat, when Mary is tired and when Mary is scheduled to get her medicine. Pepper even schedules Mary's doctor appointments for her.

Pepper also knows when Mary needs to get out of her wheelchair to use the rest room or when Mary needs to be lifted out of bed. In those cases, Pepper calls on another new friend, Robear. Robear is very strong, but very gentle. Robear helps all of the people at the nursing home when they need to be lifted out of bed or into or out of a wheelchair. Robear does this an average of 40 times per day without ever getting injured or tired. Robear takes no sick days and will never have a worker's compensation claim for a bad back.

As you have guessed by now, both Pepper and Robear are humanoid robots, and both of them exist today. Pepper, developed by Softbank Robotics has the ability to perceive emotions, and was designed to be a day-to-day companion. Pepper recognizes people's faces, interacts and moves autonomously. Pepper can analyze the look on your face, your body movements and the words that you use to learn how to respond to your mood at that point in time. Those who have interacted with Pepper are consistently surprised by what Pepper can do.

Robear was designed to look like a teddy bear so that it is not frightening to the elderly or others who need its assistance. The robot has specially engineered parts that allow it to have a gentle touch, so even though it is strong enough to lift large weights, it is also nimble enough to have the person not feel like they are being manhandled.

Right now Robear is expensive, in the $200,000 range, but like all electronics it is anticipated that the cost will come down substantially as the systems are more fully developed. Developed by Riken and Sumitomo Riko Company Limited, it was designed to address the problem of an aging population in Japan, which has a shrinking population which will be made up or more and more elderly people, with less healthcare workers to service the population.

In addition to Pepper and Robear, there are other companies developing robots to work with the elderly. One other such robot is being developed by a French company called Yumii. Yumii's companion robot goes by the name "Cutii." Cutii was created for elderly care in the home, as opposed to an institutional setting.

Cutii has vocal and facial recognition. It can provide access to a "full catalog of activities and services."

On your next trip to Disneyland, don't forget to stop by the Autopia ride, where you will find ASIMO (Advanced Step in Innovative Mobility). ASIMO was the first real humanoid type robot that has been in development since the 1980s by Honda. ASIMO has the ability to recognize faces, can detect moving objects, can distinguish sounds and react to commands, and generally interact with humans. ASIMO can also walk, dance, run, and pick things up. Obviously, all of these robots are still in development, and Honda received negative press following the March 2011 earthquake and tsunami that crippled Japan's Fukushima nuclear plant. Honda received numerous requests to send ASIMO to help with the recovery, but ASIMO wasn't designed for disaster recovery, or even evolved enough to work outside a lab or office environment. As a result, Japan relied upon robotic assistance from American companies, and later Honda announced that it was now working on a disaster recovery robot.

In 2016, Honda announced that it would work together with SoftBank, the Internet company that developed Pepper, to develop artificial intelligence products with sensors and cameras that can converse with drivers of automobiles.

Boston Dynamics, which was acquired by Google in 2013, is another leader in robotic research and development. In a surprising move, as of this writing, Google (Alphabet) has Boston Dynamics up for sale. The Alphabet public relations department reportedly thought that the company's Atlas and Spot robots were "terrifying."

Boston Dynamics developed Atlas, a two legged humanoid robot and a robotic dog named Spot. In early 2017, the company released video of its latest robot, Handle. Handle appears to be nothing like any of the previous incarnations of robots, and does not follow the concept of attempting to look somewhat human. Handle looks to be best suited for working in a warehouse. Handle has wheels where we humans have feet, can jump over hurdles and land on its wheels, is extremely stable, can lift heavy items, and can go down stairs. Handle is very dynamic and is unlike previous stationary factory robots. Although it's not quite ready for prime time, the video of Handle is impressive, and anyone interested in what a robot can do should take a good look.

The main focus of robotic development has been in the manufacturing field. While there have been large stationary manufacturing robots used for many years in sectors like the automotive industry, the newer versions of manufacturing robots are smaller and more adaptive. These robots are also designed to be less dangerous, as the larger, older robots are typically sealed off from areas where humans work due to the dangerous nature of their activities.

Rethink Robotics is a developer of industrial robots, and created Baxter in 2012. Baxter was designed to perform simple industrial tasks, such as routine production line work. Baxter is relatively small, only three feet tall, and has two arms and an animated face. Baxter is different than previous industrial robots in that it is adaptive, and does not necessarily have to do one task over and over again. Baxter can be set in a stationary position, or can be mounted on a pedestal with wheels to make it mobile. Baxter was designed so that it could be programmed easily by workers, and does not need extensive coding to program activities. It can be taught to perform tasks in minutes, unlike older versions of robots that require computer engineers to write new code for new tasks. Baxter has multiple sensors embedded in its hands and arms and consequently is designed to be safe to work in close proximity to humans. Workers can "program" Baxter for a task simply by guiding its hands through the training motions.

Three years after Baxter came on the scene, Rethink Robotics came out with Sawyer. Sawyer weighs in at only about 25% of Baxter, is more flexible, and only has one arm. Sawyer was created to perform smaller and more detailed tasks than Baxter. Rethink Robotics refers to their products as "smart, collaborative robots." Sawyer was designed to do more intricate tasks than most industrial robots. The company says that with Sawyer's "built in force sensing, you can control force for applications where delicate part insertion is critical, or use force feedback in tasks where you need to verify that parts have been seated properly."

Rethink Robots believes that these robots will support and supplement workers in the manufacturing field, not replace workers. The idea is to have the workers spend more time using their intelligence, and less time doing the repetitive and dangerous tasks that the robots can accomplish.

One of the more interesting, and some would say bizarre, companies in the robotics filed is Festo, a German industrial automation company.

Using nature as its inspiration, Festo creates robots based on animals and insects as ways to improve technology. Festo seeks to understand just how and why some creatures move and fly. The results of Festo mimicking biology are flying seagulls, hopping kangaroos, robotic ants that cooperate (BionicANTs), eMotionButterfly, AirJelly (flying jellyfish), AirPenguin, AquaPenguin, AquaRay, AirRay, AquaJelly, SmartBird, and BionicOpter (a robotic dragonfly). Festo calls its group the Bionic Learning Network, and has published technical papers with groups like the International Congress of the Aeronautical Sciences.

As funny and interesting as the unique robots seem, there is method to their madness. Festo has been a leader in pneumatics and industrial automation over a period of 50 years. The company is focused on "drive, control and gripping technology," owns about 3,000 patents, comes out with over 100 new innovations per year, and has a catalog of about 30,000 products with thousands of variants.

In a test that has similarities to the BionicANTs robots, the US Department of Defense released a video of 103 Perdix micro-drones launched from three F/A 18 Super Hornets. These micro-drones demonstrated unique ability in the form of swarming like flocks of birds or bees. The drones were not pre-programmed individuals, but a collective, where the swarm had no leader, and adapted to other drones entering and exiting the team. The drones performed a series of missions, where they surrounded a designated point on the ground, and demonstrated "one distributed brain for decision making and adapting to each other."

The Perdix drones began as a student project at MIT, and later adapted for military use by MIT Lincoln Laboratory. Most fascinating is that the commercial components used in Perdix drones are based on those used in smart phones and 3D printable parts. The test also confirmed that the drones could survive being launched from aircraft travelling at over 450 mph, and -10 degrees Celsius from high altitude. In the test, the drones simply patrolled an area and demonstrated swarming capabilities, but it seems obvious that they may be used for surveillance or even attacks, and to maintain battlefield communications.

All of the above are the best known robots and companies operating in the areas described, although substantial venture capital is being invested in robotics and new companies are forming constantly. The eventual leader in the robotics arena may be a company that does not yet exist.

CHAPTER II

TECHNOLOGICAL ACCELERATON – HOW DID WE GET HERE?

Artificial Intelligence is growing up fast, as are robots whose facial expressions can elicit empathy and make your mirror neurons quiver.

Diane Ackerman

Artificial Intelligence has seemingly burst onto the scene out of nowhere. Is that really true? Not quite. Our arrival at this point was basically a function of Moore's Law.

Many of you are aware of Moore's Law, but for everyone who is not familiar with the term, it relates to a prediction made by Gordon Moore, one of the founders of Intel. In 1965, Gordon Moore observed that the number of transistors per square inch on integrated circuits had doubled every year since their invention. He saw this as a trend, and extrapolated that computing power would double approximately every two years for at least the next ten years, with the relative cost remaining the same (or slightly lower). The only thing that surprised Gordon Moore is that this exponential trend did not continue for ten years, but has now continued for fifty years.

Most of us, yours truly included, cannot actually conceive of exponential growth, but it is best explained by the ancient chess legend. The story has many variations, but the simplification of it is that a king wished to bestow a gift upon a worth servant, who only asked that his family be well fed, so would the king simply give him one grain of rice on the first square of a chessboard, and then just double it for each successive square? No problem thought the king. Big problem, unknown to him.

The changes are tiny and incremental for a while, then something happens. There are sixty four squares on a chessboard, and for the first half, the number gets pretty big, 4,294,967,295 grains of rice to be exact, or about 100,000 kilograms of rice. While that is a lot of rice, the second half yields more than two billion times as much as on the first half of the chessboard.

It appears that the reason that computer technologies have essentially gone wild in the past few years is that the changes appeared to be so small and almost incremental for so long. A change from .001% to .002% is hardly noticeable, but a change from 400% to 800% is hard to ignore. Many refer to this as the "great acceleration," but it was always great, just not easy to identify to non-computer people. Now with the increased speed and the combined interaction of various technologies, the changes are happening on a daily basis.

Ray Kurzweil is probably the best known person who could be categorized as a "Futurist." Mr. Kurzweil coined the phrase "second half of the chessboard" which references the point where exponential growth really explodes.

So, where are we with regard to Moore's Law, and the doubling of computer power? On the second half of the chessboard. Similarly, related technology such as data storage, software development, communications bandwith, and related matters have also accelerated beyond our basic comprehension, and have gone hand and hand with computer power. While exponentials can't go on forever (physics won't let them), no one expected technological growth to reach the levels we are now seeing, and there are some legs left in the process.

Kurzweil, and others like him, also believe that once we reach the end of the ability of the presently used components, there will be a paradigm shift which will continue the forward progress. Some explain this process by saying that progress moves exponentially and levels off, at that point, great minds come up with a new solution, a new process, new components, or whatever is necessary, and then the exponential forward progress begins anew.

To make all of this even more interesting and more complicated, the growth in the areas of sensors and networks have propelled developments in computer technology forward by allowing the rapid deployment and connectivity of varying systems.

A sensor is a device or electronic component that gathers input from its surroundings. It can detect things like heat, light, sound, pressure, magnetism, radiation or motion.

At the Emerging Sensing Technology Summit held in Australia in 2016, the main themes for discussion were "optical and microwave sensors, chemical and gas sensors, biosensors, sensing materials, thick and thin film sensors, antennas, ultra low power sensors, flexible and wearable sensors, remote sensing, imaging, radar sensors, sensor modeling, extreme sensing, single chip sensors, machine olfaction for environmental sensing, and sensor networks." The applications for sensing technology appear to be endless, aviation, automotive, agriculture, health and medicine, military, and environmental among others.

Another industry group is MEMS and Sensors Industry Group, the trade association advancing microelectromechanical systems (MEMS) and sensors across global markets. This group has sponsored seven TSensor Summits between 2013 and 2015. TSensor stands for "trillion sensors." There are various forecasts and predictions for the sensor market, with one forecasting 100 trillion sensors by 2030. These predictions are derived from the forecasted growth of the Internet of Things (IoT) and mHealth, which should reach $32.5 trillion by 2025. With a GDP of $18.5 trillion for the United States in 2016, this is a massive prediction. Other industry sources note that the energy consumption related to these projections is an issue that has not been fully analyzed.

Sensors are great, but they wouldn't be much without their closest companions, networks. Without high speed networks, sensors would be lost. Fortunately, the development of sensors and networks has gone hand in hand, and the explosion in the development of both simultaneously has created opportunities and possibilities unheard of in the past.

The speed at which data is delivered is a key ingredient of the technology mix. The original 2G networks operated at 100 kilobits per second (Kbps). To keep this in perspective, downloading a song of 4 MB requires less than 40 seconds if the data speed is 1 megabits per second (Mbps). The same song requires a much longer time of five minutes to download if the data speed is 128 Kbps.

Typically, 3G networks offer download speeds from 600 Kbps up to 1.4 Mbps. Today in most areas we have 4G networks which deliver data at 5 Mbps to 12Mmbps.

As fast as those systems are, the next generation makes them look almost childish. There are companies today working on optical polymer materials systems for applications in high speed fiber-optic data communications and telecommunications that can deliver data at 100 gigabits per second (Gbps). They accomplish this by using their 25 Gbps modulator in a 4 channel architecture. They may also be releasing the same architecture using a 50Gbps modulator to deliver data at 400Gbps. This means that these systems will operate more than 1,000 times faster than the 4G networks.

With additional speed comes traffic, lots and lots of traffic. In 2005, there were approximately 500 million devices connected to the Internet. In 2016, that number reached 6.4 billion. Cisco, the networking giant, estimates that 50 billion devices will be connected to the Internet by 2020. Intel says that Cisco is wrong, that the number will be more like 200 billion, not 50 billion.

Since those projections were made, other industry participants have toned down the enthusiasm. The estimate by the CTO of Stringify, is now 30 billion connected devices by 2020, and Ericsson expects 28 billion by 2021. In addition, IHS Markit projects 30.7 billion IoT devices for 2020, Gartner projects 20.8 billion, IDC anticipates 28.1 billion.

Even the revised numbers are astonishing, and if they are even half right, the growth curve is outrageous.

One of the results of this technological growth spurt has been the growth of artificial intelligence. Keep in mind that AI is not a single technology, but a combination or collection of technologies working together, to perform specific tasks, and working at an incredible speed.

The story of Watson is a great example of this concept.

CHAPTER III

WHAT CAN ARTIFICIAL INTELLIGENCE DO FOR ME?

> *The benefits of having robots could vastly outweigh the problems.*
>
> *Rodney Brooks*

Watson, developed by IBM and named after its first CEO, Thomas J. Watson, first came into our common consciousness when it appeared on the television show *Jeopardy!* in 2011, competing against two of the show's most prolific winners. Watson answered questions posed in natural language and competed for a $1 million prize. Watson beat its two human competitors, and has gone on to bigger and better things.

Actually, Watson was not IBM's first foray into demonstrating the abilities of a computer versus a human. In 1997, a computer called "Deep Blue" defeated Garry Gasparov to become the first computer to beat a reigning world chess champion in a match under standard chess tournament time controls. Gasparov had beaten Deep Blue in a tournament in 1996, before it was upgraded.

These early activities of IBM drew attention to the fact that computers were getting better and better, and basically demonstrated the effects of Moore's Law. What happened after Watson's appearance on *Jeopardy!* however, is much more earth shattering.

In early 2013, Memorial Sloan-Kettering Cancer Center in New York City announced a joint venture with IBM to develop what it called "Watson for Oncology" sometimes referred to as WFO. According to Medical Marketing & Media, WFO is "a cognitive computing system that can analyze large volumes of data including medical literature, patient health records, and clinical trials, to offer personalized, evidence-based treatment recommendations for cancer patients."

The team at Memorial Sloan-Kettering in essence took Watson to medical school. Watson began "studying medicine" in 2011, and has not stopped since. By most accounts, Watson has ingested over 15 million pages of medical information, including 200 textbooks and 500 medical journals.

Depending upon what sources you believe, there are anywhere from 160,000 to 700,000 research papers published every year in the medical world. As far as we can determine, human researchers read about 200 published research papers each year. It is virtually impossible for any human to keep up with the massive amount of information flooding the medical world each year. Watson reads it all, and retains the information for future use. Since it is technically impossible for any doctor to read, review and retain all of the medical journal articles per year, Watson is designed to assist doctors in reviewing all relevant information to assist in diagnosis and to help design personalized treatments. It has been said that by 2016 Watson had sifted through 20 million cancer research papers.

As we all know, cancer is a massive problem for the entire population. According to the National Cancer Institute, there were approximately 1.7 million new cases of cancer in 2016, and nearly 600,000 deaths. With about 40% of Americans being diagnosed with cancer during their lifetime, it requires new and better approaches to diagnosis and treatment, such as Watson.

At Memorial Sloan-Kettering, they believe that they "can decrease the amount of time it takes for the latest research and evidence to influence clinical practice across the broader oncology community, help physicians synthesize available information, and improve patient care." Along with clinical trials and contributing to oncology organizations, the hospital treats over 130,000 patients who are afflicted with cancer.

In recent years, Watson has been fed not just technical medical journal type information, but also medical information such as treatment guidelines, electronic medical record data, notes from physicians and nurses (in many cases hand-written), clinical studies, radiological data, image identification, and patient information. The concept is to have Watson assess and extract large amounts of data, both structured and unstructured, from all of this input using natural language processing and machine learning. Watson is also intuitive in that when it finds inconsistencies in patient files, like a lab report that does not coincide with notes from a doctor, it flags this information for further review by the physicians.

Watson has capabilities that are akin to human skills, in that it can read natural language, evaluate cases, evolve with machine learning, and rapidly process massive volumes of data. Over the years, Watson has been given detailed patient information, without identifying the name of any patient, so that it can "understand" the dynamics of a patient's condition, and can apply its knowledge to specific situations.

A prime example of the use of Watson involved working with Doctors at the University of North Carolina School of Medicine. These Doctors provided Watson with the records of 1,000 cancer patients, and Watson provided treatment plans that concurred with oncologists' actual recommendations in 99% of cases. In addition, since Watson has been fed all of the latest cancer research, it was able to provide additional treatment options missed by its human counterparts in 30% of the cases.

In situations involving rare cancer conditions, Watson has proved itself to be a helpful assistant to doctors. For example, the University of Tokyo reported that Watson had correctly identified a rare form of cancer in a 60 year old female patient. The patient had been diagnosed with acute myeloid leukemia, a different form of cancer than she actually had, and had been receiving treatment for a number of months. The treatment protocol was ineffective and the physicians did not understand why the patient was not responding as anticipated. The doctors then turned to Watson and after providing Watson with all of the medical records of the patient, her genetic profile and shifting through millions of cancer research papers, Watson correctly determined the actual disease and recommend a new treatment. What is even more amazing than the outcome, is that it reportedly only took Watson ten minutes to come up with the diagnosis and the new treatment plan. The last report on the patient was that she was responding well to the new treatment.

In another success story, Watson was employed to consult on the case of a 37 year old software engineer in India who was diagnosed with a rare and aggressive form of breast cancer. Based on the specific information fed to Watson about this patient, it took 60 seconds for alternative treatment plans to be generated by Watson. According to reports, through a drug treatment protocol and some surgery, both of the patient's breasts were saved.

These two examples are demonstrative of the remarks of Dr. Craig Thompson, President and CEO at Memorial Sloan-Kettering, who commented at an IBM investor briefing in 2016 that 20% of cancer patients in the United States are misdiagnosed. If Watson can assist doctors with just a portion of that problem, it will be well worth its investment.

Watson has now been incorporated into hospitals around the world, and at last count was being used in over 30 major hospitals. Along with hospitals in the United States, like Cleveland Clinic Lerner College of Medicine of Case Western Reserve University, and Westmed Medical Group on New York, Watson is being used overseas in Manipal Hospitals, a large hospital chain in India.

In early 2017, IBM and Jupiter Medical Center announced that Watson for Oncology trained by Memorial Sloan-Kettering would be adopted by Jupiter

Medical Center, making it the first community hospital in the United States to adopt Watson.

As the first community hospital to adopt Watson, the Jupiter Medical Center is in a unique position, since to date Watson has only been available in a few, very large institutions and cancer clinics. By having Watson available on a local level at a community hospital, it represents the fact that this kind of approach will soon be widespread and available to cancer patients in smaller communities without having to travel to major metropolitan centers for treatment and analysis. Jupiter Medical Center is ahead of the pack in implementing Watson, and is an example of a regional leader in integrating technology like Watson to provide state of the art healthcare.

Jupiter Medical Center, located in Jupiter, Florida, is a not for profit 327 bed regional medical center. The hospital was founded in 1979, and has approximately 1,500 team members, 575 physicians and 640 volunteers. While the hospital provides a broad range of services, it provides specialty concentrations in cardiology, oncology, imaging, orthopedics, digestive health, emergency services, lung and thoracic services, women's health weight management and men's health.

At the time of this writing, Watson has been trained in six types of cancer, and there are plans to add eight more cancers to its system. In conjunction with Watson's cancer training, IBM has added Watson for Genomics to train Watson in genomic analysis to assist in fighting cancer. In addition, as part of IBM's focus on Watson, the company has spent approximately $4 billion as of 2016 to purchase companies that own massive amounts of medical data which are being added to Watson's widening array of medical information.

It is also best to remember that Watson is not a stand alone system to replace a doctor. Watson is more like a very fast, very efficient physician's assistant, who is there to make sure the doctor has every tool that she or he needs to provide the best patient care. In a very short time, Watson can provide a doctor with all of the treatment options, and share the evidence it used to reach those options so that the doctor and patient can be fully informed and can reach a joint decision on treatment. Some have said that it is almost like getting a second opinion.

In 2014, IBM created a business group known as the IBM Watson Group, and has been working with companies in the pharmaceutical industry, publishing and biotechnology areas. Watson has also been involved in weather forecasting and tax preparation. At present, Watson is an example of AI working in conjunction with humans to assist them in their work, not to replace them.

Of course, not everyone is enthralled with Watson, and when you are dealing in the technology industry, there are always varying opinions.

Chamath Palihapitiya is a former Facebook executive who has become a venture capital investor through his company, Social Capital. Generally, his two favorite companies appear to be Amazon and Alphabet (Google). Many technology industry insiders believe that Amazon and Alphabet are ahead of everyone else that is working on artificial intelligence systems, including IBM. In addition, there are many companies working on artificial intelligence applications, as demonstrated by the fact that more than 300 companies working in the artificial intelligence sector raised seed or early stage venture financing in 2016, and the sector is growing each year.

Palihapitiya was recently criticized for his comment in May 2017 on CNBC's Closing Bell that "Watson is a joke, just to be completely honest." He later walked back his statement and said he should have been more careful with his words. His comments were in reference to one of his venture capital investments, which he said he would love to see it in a head-to-head comparison with IBM. The company in question is Syapse, and its website describes its platform like this:

"Syapse Precision Medicine Platform is a comprehensive software suite used by leading health systems to support the clinical implementation of precision medicine in oncology and other service lines. This category-defining platform enables clinical and genomic data integration, decision support, care coordination, and quality improvement at point of care."

As for IBM, their written response to Palihapitiya's joke comment was this:

"Watson is not a consumer gadget but the A.I. platform for real business. Watson is in clinical use in the U.S. and 5 other countries. It has been trained on 6 types of cancers with plans to add 8 more this year. Beyond oncology, Watson is in use by nearly half of the top 25 life sciences companies, major manufacturers for IoT applications, retail and financial services firms, and partners like GM, H&R Block and SalesForce.com. Does any serious person consider saving lives, enhancing customer service and driving business innovation a joke?"

We certainly don't.

CHAPTER IV

PREDICTIONS ABOUT THE EFFECT OF AI & ROBOTS

If robots are to clean our homes, they'll have to do it better than a person.

James Dyson

Are we at the most dangerous moment in the history of humanity as Stephen Hawking has said? Or, are we at the pinnacle of positive achievement as humans?

These comments are brought about by the belief by Hawking that we have reached the point in time where the ability of the human mind and the ability of computers are equal, where there is no difference in their abilities. What kind of future lies ahead?

In a 1942 short story called "Runaround" by science fiction writer Issac Asimov, he set forth what he referred to as the Three Laws of Robotics:

1. "A robot may not injure a human being or, through inaction, allow a human being to come to harm.

2. A robot must obey the orders given it by human beings except where such orders would conflict with the First Law.
3. A robot must protect its own existence as long as such protection does not conflict with the First or Second Laws."

Over the years, Asimov and other science fiction writers made variations to these laws, but the concept is basically the same, ie: that no harm should come to humanity by a robot.

Will those laws apply to the future? Will the harm, even if not physical harm, be in the form of unemployment to humanity? Will humans be able to control robots who can evolve much faster than a biological species?

By anyone's evaluation, Stephen Hawking is a brilliant man. Hawking's friend, Elon Musk, founder of SpaceX, Tesla, Inc. and other companies, is also fearful of artificial intelligence. Both of these men, along with various Google researchers, who are the ones leading the way in development of artificial intelligence, have expressed concern about the creation of something that can match or surpass human intelligence. What they really fear is the law of unintended consequences, where we intended for this to be beneficial to society, and instead we end up with a "Skynet" system that wants to eradicate humanity because it is inefficient. In essence, it might see us as a useless part or a computer virus.

Another expert whose name is familiar and almost synonymous with AI is Ray Kurzweil, the founder of Singularity University, and the author of *The Singularity is Near*. Kurzweil has made many predictions about the future over the years, and appears to have an excellent track record in seeing what is next. He predicts that in 2029, artificially intelligent computers will do everything better than humans. Perhaps both Hawking and Kurzweil are right. Perhaps today the ability of computers and the ability of the human mind are equal, but it will take until 2029 for the distribution system to integrate computers into everything we touch, so that they can do everything for us all in 2029.

When we talk about AI, we need to think about the three basic skills that AI is designed to perform. Those skills are seeing (reading), writing, and integrating knowledge. Those skills are part of what is now known as machine learning.

In the late 1960s, the US Postal Service (then known as the U.S. Post Office) began to experiment with sorting mail by machine. Having family members who have done this, I know that a post office employee was required to take a test, which involved the individual memorizing addresses and zip codes and sorting the mail by hand, by placing letters into appropriately labelled pigeonholes. If you worked in a post office in the 1960s, one of your skills was knowing where local addresses were located and which mail delivery truck they went on, so that when you sorted the mail by hand, each piece got to the right mailman for delivery to your door. Interestingly enough, in London in those days, the British mail system was so efficient that mail was delivered twice a day, morning and evening. If you posted a letter in the morning, it would be delivered in London that afternoon. All done by hand, and all done by humans, no machines at all.

The postal service began installing massive machines which were programmed to sort the mail, which started with package sorting. Huge machines with long conveyor belts moved the packages to the right delivery people increasing efficiency. The evolution of this process was slow, but eventually it became normal for the mail to be sorted by machines.

Over the past five years, the ability of machines to understand and differentiate has exploded. In a well-documented test in 2001, thousands of traffic signs were used in a variety of circumstances, including partially obscured, in glaring sunlight, at great distances and similar situations. The human error rate in that test was 1.2%, with the machine recognition beating humans with a half of one percent error rate. The following year, a competition was held using thousands of different images (the "ImageNet Competition"). The competition was won by a computer that was able to categorize random objects better than any of the human competitors.

In the legal profession, it has become commonplace to use "e-discovery software" to search through thousands and thousands of documents searching for specific terms or even certain patterns. All of this work used to be done by hand, by numerous paralegals combing through massive amounts of documentation. The time and cost of this work has been substantially reduced. Today software of this type is used to manage and analyze electronic data during investigations and litigation by all of the major law firms. Software of this type is used to analyze emails, documents, presentations, databases, voicemail, audio and video files, social media, and web sites, and any other electronic data.

The internet has created a need for content of all types. Today, much content is not written by people, but by machines. Having reviewed articles written by machines and compared them to articles written by humans, I can attest to the fact that it is becoming more and more difficult to tell the difference. In the investment world, some research reports are being written by machines, after the machines have been given orders to search for certain types of information. A company called Narrative Science invented "Quill" which is a "an advanced natural language generation technology." Feed Quill the data on a company, and it will generate a report written in English which will discuss the upside, downside and other risks and unique attributes of a company.

The third skill of machine learning, and the most important one, is the area of integrating knowledge. This is the area where a computer learns to pull together all of this information, and actually reach a logical conclusion based on all of the information in its system. As we have explained in a prior chapter, this is a skill demonstrated by Watson.

So what will all of this technology do to the workplace as far as jobs are concerned? Peter Norvig, a well known artificial intelligence scientist is concerned about the mass elimination of jobs that he believes is on its way. His view is that while we have seen a lot of change, the disruption will continue. While he does not hold the same negative view of Hawking and Musk about "killer robots," he is worried about the impact of expanding AI and its ability to have a negative impact on the job market. Just as most scientists in this area, he is not sure if the impact will be minimal, or if it will consume half of all jobs over the next decade. He reminds people that while it is easy to see what jobs might disappear, none of us may be able to comprehend what new jobs may emerge. The issue appears to be at what speed will all of this occur. Will there be time for many people to adapt or will the changes come so rapidly that large portions of society will find themselves out of work literally overnight? Part of his advice is that there will always be something to do, and everyone needs to be very alert and aware of what is happening in their industry, so that they are not totally shocked by the changes that are inevitable.

Until computers started demonstrating the ability of integrating knowledge and actually suggesting conclusions and answers, everyone knew that routine tasks were fair game to be replaced by automation. After AI starting showing its abilities in nonroutine and complex matters, that perception changed. When we combine machine learning and advanced robotics we have computers that are answering natural language questions, driving trucks, seeing patterns in writing that humans can't recognize and performing tasks beyond the ability of people.

In 2013, Frey and Osborne published a study known as *"The Future of Employment."* This study, which has been widely recognized, predicted that there was a greater than 70% possibility that approximately 47 percent of all jobs in the United States are at high risk of being automated over the next few decades. This estimate is based on advances in artificial intelligence and the continuing explosion in computer robotics.

Similarly, Stuart W. Elliott wrote an article called *Anticipating a Luddite Revival* in 2014, which discussed his study of AI and robotics and their effect on the economy. Elliott's investigatory study "sampled articles from two journals, AI Magazine and IEEE Robotics & Automation Magazine, over a period of 10 years, from 2003 to 2012." Elliott was confronted with a massive amount of information, and to assist those who are not so technically oriented and to assist in the understanding of the subject matter, he separated the capabilities into four general areas: language, reasoning, vision, and movement.

In the language area, there were four specific aspects of language that were studied: understanding speech, speaking, reading, and writing. Generally speaking, the articles from the first five years (2003-2007) were not quite as complex as those from the second five years (2008-2012).

In the area of reasoning capabilities, different aspects of reasoning included recognizing that a problem exists, applying general rules to solve a problem, and developing new rules or conclusions. Elliott noted that the demonstration of common sense type reasoning was not evident in articles from the first five years, but that some systems analyzed as part of the later five years of articles demonstrated more broad and flexible reasoning.

With regard to vision capabilities, the articles discussed systems that recognized objects and different features of those objects, and this vision capability included identifying the position of objects in space. Again, the earlier articles discussed systems that had capabilities that were not as advanced as the later articles, indicating and perhaps demonstrating the changes due to Moore's Law. All of the systems that were studied involved identifying diverse objects and recognizing features of the identified objects, including their location and movement.

As for the movement capabilities, systems reviewed involved spatial orientation, coordination, movement control, and body equilibrium. Elliott noted that many of the systems integrated movement capabilities with one or more of language, reasoning and vision capabilities.

After identifying these capabilities, Elliott then compared them to the skills required in different occupations. Elliott used the U.S. Department of Labor's O*NET system, which provides ratings for hundreds of occupations.

Elliott's study "suggests that there is the technological potential for a massive transformation in the labor market over the next few decades." Elliott also stated that "In principle, there is no problem with imagining a transformation in the labor market that substitutes technology for workers for 80% of current jobs and then expands employment in the remaining 20% to absorb the entire labor force."

Elliott reflected upon the fact that the United States changed from an agrarian economy to an industrial economy. This move from 80% of the population being employed in agriculture to just a few percent of workers, however, took place over a century and a half, while the current evaluation of the AI and robotics movement is anticipated to occur over just a few decades. Can we adapt that quickly?

Another analysis came from the McKinsey Global Institute, whose researchers did not focus on entire job replacement, but instead they evaluated approximately

2,000 individual activities that were part of various occupations. The overall estimate of the Institute was that as many as 45% of all activities performed by individuals in their work could be automated with existing technologies. Those activities represent about $2 trillion per year in wages. More specifically, the analysis of the Institute was that 5% of all jobs could be totally automated and eliminated, and that about 60% of all jobs could have about 30% of their tasks automated with today's technology. With regard to the 45% of activities that could be automated today, McKinsey estimates that the percentage could rise to 58% "if natural-language processing were to reach the median level of human performance." *McKinsey Global Institute, 2015, Four Fundamentals of Workplace Automation.*

McKinsey also indicated that automation is not limited to low paying, low skilled jobs, but that "a significant amount of activity" could be automated in the highest paying jobs like physicians, senior executives and chief executive officers.

One of the most important comments from the McKinsey study was this:

"When we modeled the potential of automation to transform business processes across several industries, we found that the benefits (ranging from increased output to higher quality and improved reliability, as well as the potential to perform some tasks at superhuman levels) typically are between three and ten times the cost. The magnitude of those benefits suggests that the ability to staff, manage, and lead increasingly automated organizations will become an important competitive differentiator."

This statement becomes more and more important as we continue to discuss the future of capitalism.

The McKinsey study determined that the future focus will be less about completely replacing certain jobs, and more about automating particular portions of jobs. In reaching this conclusion, they focused on four insights: the automation of activities, the redefinition of jobs and processes, the impact on high wage occupations, and the future of creativity and meaning.

This fourth aspect, creativity and meaning, is clearly the most important area of focus for the future of capitalism. The McKinsey study points out that creativity and sensing emotions are difficult to automate, and are a basic human trait. As important as they are, only a very small portion of activities of workers involve these capabilities. Only 4% of activities involve creativity "at a median human level of performance" and about 29% of work activities require "a median human level of performance" in sensing emotion. This may be a prime example of workers spending time and energy for routine, automatable tasks, and simply not having adequate time to indulge in the uniquely human forms of creativity in the workplace. With workers freed of mundane, repeatable, routine tasks, it is possible that productivity would increase accordingly.

World Economic Forum

In 2016, the World Economic Forum published a Global Challenge Insight Report titled *The Future of Jobs – Employment, Skills and Workforce Strategy for the Fourth Industrial Revolution*. This insightful and interesting work is 167 pages full of data, concepts and useful information. While there is no way to replicate the entire report here, we will address some of its comments, and refer you to the entire report for more detailed information.

This fourth industrial revolution is being driven by information technologies and advances in computing. Industry leaders polled for this report identified the areas where they believed change would be coming, and estimated time frames for such changes. Between 2018 and 2020, these leaders see further development in advanced robots with enhanced dexterity, senses, and intelligence, particularly in the transportation field. The also see the potential to automate many knowledge-worker tasks which have previously been viewed as impossible for machines, all due to artificial intelligence, machine learning and natural user interfaces through voice recognition. In addition, technological advances in 3-D printing will revolutionize many manufacturing processes, and technological advances in life sciences will affect medicine and agriculture with new pharmaceuticals, polymers, biofuels and other new materials.

Of course, educational institutions, businesses, governments, and nonprofits will need to refocus efforts on helping workers obtain the skills needed to succeed in the new economy. The World Economic Forum report states that one popular estimate is that 65% of children presently entering primary schools will end up working in new job types that do not exist today. It will be difficult, if not impossible, to train people in advance for jobs that may develop later and which are undetermined today. The best that we can do is try to prepare young people broadly so that they learn technical skills as well as creative and social skills, so that they may adapt when the time comes.

Since many of the changes anticipated will take place over a few decades, as opposed to a century and a half, education needs to become more of a lifelong learning process for all workers, so that future skills can be met. Obviously, this is much easier said than done, but one focus needs to be in continued training offered through the existing workplace environment, and through other incentives such as tax credits for additional training.

Will it ever be possible for AI to exhibit what we now see as uniquely human skills? While many researchers believe that to be the case, some of those skills requiring creativity may be beyond the reach of machines. At least it appears that will be the case for a while.

Is there a guarantee that new technology will be deployed? Decisions to use or not use technology depend upon a cost/benefit analysis. In all of the analysis that has been done by researchers regarding the potential of technology to accomplish tasks done by humans, it may be impossible for them to quantify the human characteristics that go into some of these jobs, and thus their analysis may be flawed.

Of course, we do not know how quickly changes will come, and the speed of change will depend upon the interconnection and interplay between governments, organizations, policies, national cultures, and society in general. Market demand and acceptance by society of automation practices will also drive the pace at which technology is implemented. Remember, people don't always enjoy talking to a machine, and when you multiply the tasks to be accomplished by AI, some of them may not be market friendly, and thus unacceptable. Much of this goes to the comment that just because you can do something doesn't mean you will. Today automation can replace many tasks done by people, but will it?

So, part of the question is will all of this AI and robotics stimulate economic growth, and at what cost? Will companies become more productive, but employ less people?

The focus we believe should be on better living standards for the population, and how that may be accomplished. In this regard, the findings of Robert Solow are of importance to us all. Mr. Solow developed an economic theory of growth, which stated that most of the increases in human living standards have come not from working more hours, and not from using more capital or other resources, but from improved productivity. Improved productivity is an increase in the efficiency of production as defined by the ratio of output to input. His determination was that productivity growth comes from new technologies and new techniques of production and distribution. (R.M. Solow, 1959, A contribution to the theory of economic growth, *Quarterly Journal of Economics* 70(1):65-94, doi: 10.2307/1884513)

You have undoubtedly heard that to get ahead you don't need to work harder, you need to work smarter. Working smarter means using techniques, technologies or whatever else to get more done with less effort, ie: increased productivity. Therefore, according to Solow's theory, advances in technology, while not initially beneficial to all people simultaneously, will benefit society as a whole by increasing productivity.

We should all keep in mind that when it comes to the future, the experts probably don't really know any more than you or I do.

CHAPTER V

THE NEW WORLD OF EDUCATION

Online learning is not the next big thing, it is the now big thing.

Donna J. Abernathy

As the world around us changes, it was inevitable that concepts relating to education would evolve. In response to our technical environment, various new approaches to education have emerged. Because we must all prepare for the future, venture capital companies have focused on funding different educational platforms. This area has now come to be known as "EdTech."

As far back as 1995, a company was founded called Lynda.com, offering online courses and tutorials. Early on, Lynda Weidman saw a need for online courses. The result is that there are now thousands of videos on the site, and through its subscription service, there are approximately 2,000,000 members. This company was so successful that LinkedIn Corporation bought it for a reported $1.5 billion in 2015. For what it costs to be a member, its seems like a very reasonable opportunity. As of this writing, Lynda.com charges $25 per month for their basic service or $37.50 per month for the premium service. If you pay annually, the fee is $250 per year for basic and $375 per year for premium. Courses vary from basic to advanced, and the quality is typically reported to be excellent. It has certainly become one of the go to places for online education, particularly in the technology field.

One of the most recognized EdTech companies is Coursera. Coursera uses a different approach than Lynda. Coursera works with universities to provide what it calls free massive open online courses (MOOCs). The company offers a wide range of courses in various subjects as diverse as engineering and the arts. If its courses are free, how does it make money as a for-profit education company? By offering verified certifications that students list on their resumes to let employers know that they are competent in a particular area. Coursera has by all accounts been wildly successful, with over 1700 courses and over 23 million registered users. The company even has an online MBA program that is operated in conjunction with the University of Illinois.

Coursera has about 150 partners, including the University of Pennsylvania, Ohio State University, University of Virginia, Yale University, the University of Washington, the University of Michigan, Vanderbilt University, the University of Illinois, and numerous other American and foreign universities. In the past few years, Coursera has increased its "Specializations" offerings, which are targeted courses designed to build high demand skills like data science.

In 2012, Stanford University professor Sebastian Thrun launched Udacity with two partners. The prior year, the professor had offered a free online course in artificial intelligence, and after over 160,000 students from around the world signed up to take the course, he knew he had hit upon something big. When it started out, the company offered somewhat traditional university style courses, but in recent times the company's focus has shifted to more specific courses in the technology field. Udacity also offers what it calls "nanodegrees," which are courses of study that can be completed in less than a year, and typically in six to nine months. In an effort to get students qualified quickly and affordably for technology jobs, Udacity partnered with AT&T and the Georgia Institute of Technology in 2014 to create the first "massive online open degree" in computer science. Through this course, a student can obtain a master's degree at a cost of approximately $7,000.

Another important company that uses a subscription based model is Pluralsight, which has probably raised more venture capital money than any other educational online learning company. The focus of Pluralsight is technical education. When the company was originally founded in 2004, it was primarily a classroom training company that sent instructors to businesses for tech training. It moved to the online video format in 2007. Today the company has over 1,000 expert instructors and offers over 5,000 courses. Authors of courses are paid a royalty, which is based upon how frequently their videos are viewed. The company teaches courses relating to technology and its application in various fields, such as architecture and construction, manufacturing and design, business management and communications, and project management, along with core courses in cyber security, data analytics, engineering and science.

While many companies have followed either the subscription based format or the free MOOC format, two companies have employed a different business model. Udemy is a company that provides online learning through a platform that allows anyone to create their own course, which can either be offered for free or for a fee. Through the Udemy platform a user can not only create a course, but can also promote it and can earn income from student tuition charges. Udemy provides the tools and the marketplace, and as of 2017 had about 45,000 course offerings. Udemy is interesting because it does not just focus on technical courses, but provides a very wide variety of topics and subjects, some very broad and some extremely narrow.

The other online company with a different approach is Knewton. Knewton bills itself as an "adaptive learning technology provider." The company is focused on personalizing educational content for each student individually. By analyzing a particular student's performance data, it identifies the student's strengths and weaknesses. Knewton wants to bring personalized education to the world.

One of Knewton's original investors and partners, Pearson, announced in May 2017 that it was developing its own in-house adaptive learning capability. Pearson, founded in 1844, is a British-owned education publishing and assessment service to schools and corporations. According to Wikipedia, Pearson owns educational media brands including Addison–Wesley, Peachpit,

Prentice Hall, eCollege, Longman, Poptropica, Scott Foresman, and others.

According to edsurge.com, "Knewton also provides the adaptive technology backbone to other companies and traditional publishers—including Houghton Mifflin Harcourt and others that directly compete with Pearson. Through these partnerships, Knewton tags digital content with metadata and maps them onto a learning path. As students work through the materials, Knewton's algorithms detect where they are struggling and surface the appropriate content from the publisher."

As you can see from this brief introduction to the EdTech arena, anyone anywhere in the world that has an internet connection, now has the ability to learn any technical skill required today for any job, and in most cases, they can learn it for free. Never in the history of the world has this ever been possible before. Consequently, it is difficult to believe that anyone who desperately needs or wants a technical job cannot obtain the skills needed. All they need is time and dedication, since the opportunity to learn is now widely available. Those hungry for learning, in America and elsewhere, are doing just that, they are getting these skills online.

CHAPTER VI

WE HAVE BEEN HERE BEFORE

The Sky is Falling! The Sky is Falling!

Chicken Little

According to the President's Report, referenced earlier, AI could affect as little as 9 percent of jobs over the next decade or two, or it could take away nearly half of all jobs. Similar comments have been made by industry sources, and people in the transportation and food services industries fear that they will be first, and will be the hardest hit.

Before we all panic, it is best to take a historical view of what is happening. How many times have you heard that "If this trend continues, the result will be disaster. We are going off the cliff." The subject can be just about any problem, but the pattern of reaction is the same. People take a current trend and extrapolate it into the future as the basis for their gloom and doom predictions. The fundamental problem with most predictions of any kind is that they make a very serious false assumption: that things will go on as they are. People who make these assumptions overlook the very basics of economics and human nature: that given any problem, people have the insight to see the incentives, and react accordingly. Solutions magically appear.

Life After Peak Oil

A petroleum geologist, M. King Hubbert observed in the 1950s that once you extract half the oil from a given field, production begins to decline. Based on that very basic original observation, he calculated what he believed to be the point at which the United States would reach its peak with regard to production of oil. He predicted that the United States would reach peak oil in the early 1970s. Needless to say, Mr. Hubbard's conclusions were not welcome with open arms by the oil industry. As we all know, in 1970, the United States was producing approximately 9 million barrels of oil per day, and thereafter began to decline in production.

A few years ago, the buzz word was "peak oil." It nearly created chaos. Let's go back a few years and look at the context.

In 2005 the U.S. Department of Energy funded and published a report titled "Peaking of World Oil Production: Impacts, Mitigation, & Risk Management." Authored by Robert L. Hirsch, Roger Bezdek and Robert Wendling, the report is an assessment requested by the U.S. Department of Energy, National Energy Technology Laboratory.

This report has come to be known as the Hirsch Report. Robert L. Hirsch is a senior energy program adviser at the private scientific and military company, Science Applications International Corporation (SAIC).

SAIC is involved in work in the areas of defense and geopolitical issues.

"World oil peaking is going to happen," the report says. The effects and the economic impact on the United States were of great concern to the authors. "The development of the U.S. economy and lifestyle has been fundamentally shaped by the availability of abundant, low-cost oil. Oil scarcity and several-fold oil price increases due to world oil production peaking could have dramatic impacts ... the economic loss to the United States could be measured on a trillion-dollar scale," the report says.

The Hirsch Report stated that when world oil peaking occurs, it will likely be "abrupt and revolutionary." This situation presents the world and the U.S. with an "unprecedented risk management problem." The report states: "As peaking is approached, liquid fuel prices and price volatility will increase dramatically, and, without timely mitigation, the economic, social, and political costs will be unprecedented."

Time and cost are significant factors in this dilemma. "Dealing with world oil production peaking will be extremely complex, involve literally trillions of dollars and require many years of intense effort." The authors stressed that the problem is not a temporary one, and that any past "energy crisis" experience will provide "relatively little guidance." They instruct us that this problem deserves "immediate, serious attention."

The authors of the report believed that intervention of governments would be required to maintain some semblance of order. They even stated a need to exclude public debate and environmental concerns from the process. They say this was needed to speed up decision-making.

According to the Hirsch Report, mitigation efforts would require substantial time. A 20 year time frame would be required to transition without substantial impacts. A 10 year rush transition with moderate impacts would be possible with extraordinary efforts from governments, industry, and consumers. Late initiation of mitigation may result in severe consequences. It is a matter of risk management since mitigating action must come before the peak. Without massive mitigation more than a decade before the fact, the problem will be pervasive and will not be temporary. "The world has never faced a problem like this."

Another report sounding the alarm was issued in 2010. This report was issued by the U.S. Joint Forces Command, and is known as the "Joint Operating Environment", or "JOE." In describing the JOE Report, the introduction states that "It provides a perspective on future trends, shocks, contexts, and implications for future joint force commanders and other leaders and professionals in the national security field." JOE Reports are issued annually.

The JOE Report is a 74-page detailed report and it covers areas such as energy, war, climate change, globalization, and natural disasters to food, water and urbanization.

The most significant statement in the report is that by 2012, surplus oil production capacity could entirely disappear, and as early as 2015, the shortfall in output could reach nearly 10 million barrels per day. The report states that "the discovery rate for new petroleum and gas fields over the past two decades provides little reason for optimism that future efforts will find major new fields."

The report places an emphasis on the anticipated increased need for oil, coupled with the obvious inability to provide increased production. By the 2030s, demand was estimated to be nearly 50% greater than today. "To meet that demand, even assuming more effective conservation measures, the world would need to add roughly the equivalent of Saudi Arabia's current energy production every seven years."

What happened? With new oil discoveries, advanced oil drilling techniques and automotive innovations, the potential disaster was averted without government intervention or interference. Yes, there is life after peak oil, and there were no riots, gasoline hoarding problems, volatile price increases or any of the other predicted outcomes.

Another Crisis Averted - Y2K

A funny thing happened on the way to the future. Designers of computer programs in the 1970s, 1980s and even into the 1990s forgot that the next century did not begin with 19. Fear began to grip the nation's industries, the military, the government and anyone who owned a computer. What would happen after midnight on December 31, 1999? Would the world end? Would every computer in the world crash? Would all computers cease to function? Would the banking system explode? Would electric grids fail? The media had a field day and was loving all of the disaster scenarios that it could dream up.

My personal experience with Y2K took place on a trip to the northwestern states in late 1999. Going into a grocery store, I was quite surprised to see a very large section of the store devoted to disaster preparation planning. You could buy large water tanks to store water in case the system did not work. You could buy lots and lots of batteries to listen to any potential news on your battery operated radios (also for sale). Need food in case of a run on the agricultural products? No problem, we have lots of food that can be stored for years. Sleeping bags (I guess in case your electric blanket did not work), camping stoves, firewood, and everything else you can imagine. I didn't buy anything.

Industry and governments being aware of the potential issue, checked and if and when necessary, upgraded their systems, and the year 2000 came and went in peace. No crashes, no chaos, no run on the banks.

The Large Hadron Collider

According to nuclear industry sources, the Large Hadron Collider (LHC) is the world's largest and most powerful particle collider, most complex experimental facility ever built, and the largest single machine in the world. The LHC was built by the European Organization for Nuclear Research (CERN) between 1998 and 2008. Over 10,000 scientists and engineers from over 100 countries worked on the project. The goal of the LHC was to let physicists test the predictions of different theories of particle physics, including measuring the properties of the Higgs boson and searching for the large family of new particles predicted by supersymmetric theories, as well as other unsolved questions of physics.

Many crises scenarios were discussed as the LHC was being constructed. Some were sure that when the Large Hadron Collider powered up for the first time in 2008, it had the potential to open a black hole that would consume the entire world. Once again the media loved it, and played up every disaster concept they could think of to sell more newspapers. One fantastic headline, from Britain's Sun newspaper read "End of the World Due in 9 Days." Suddenly, everyone was very interested in the topic of subatomic particle research.

A German chemist even filed a lawsuit against CERN with the European Court of Human Rights, claiming a black hole scenario would violate the right to life of European citizens and pose a threat to the rule of law.

The LHC was tested in 2008, and has now operated since 2009. Although predicted by many, no black holes have opened, nor has a new big bang destroyed the Earth.

The Great Horse Manure Crisis of 1894

When you think all is lost and that we are doomed, always think about my favorite disaster in the making, the Great Horse Manure Crisis of 1894.

In 1894 all of the major cities had the same problem. None worse than London, which was then the largest city in the world, and was still rapidly growing. All of the goods and services were delivered by horse drawn vehicles. All personal transportation in cabs and busses were horse drawn. London had over 100,000 horses in the city on a daily basis, walking and running in the streets of the city, and depositing manure along the way. In the streets of New York it was estimated at 2.5 million pounds of horse manure was left in the streets every day. Pundits of the day said that urban civilization was doomed. In 1898 the first international urban-planning conference began in New York City. The conference was scheduled to last for ten days. The participants ended the conference after three days since no one could come up with a solution to the horse problem. They saw that as cities grew, more and more horses would be needed, which meant more stables on valuable land, more land would be needed to grow hay, less land would be available to grow food for people, and not one delegate could think of a solution to the horse manure problem. The London Times predicted that in fifty years, every street in London would be covered in nine feet of horse manure.

The last time I was in London, I did not see any horse manure. What happened? A guy named Henry Ford happened, and all the horse manure went away.

Remember, this is America, and do not underestimate the resourcefulness of the problem solvers among us.

CHAPTER VII
GREAT COMPANIES

Being a great place to work is the difference between being a good company and a great company.

Brian Kristofek,
President and CEO, Upshot

The year was 1925, and Florida's first major real estate boom was in full swing. Real estate sometimes fuels the hopes and dreams of young men, and George W. Jenkins was no exception to the rule. George, who was born in 1907, lived in the small town of Harris, Georgia, but just knew that his fortune was waiting for him in the booming real estate of sunny Florida. In some ways he was right, but there was a round about way that it occurred.

George saw great potential in the Tampa area, and moved there with real estate on his mind. Being all of 17, and having no money of his own, George first had to figure out how to feed himself and exist until he could get his opportunity. Since he had worked in his father's general store back in Georgia, he was able to convince a grocery store to hire him as a stock clerk. George went to work for Piggly Wiggly, and soon proved himself to be such an asset to the store that he was promoted to manager shortly thereafter. Piggly Wiggly was a chain of small grocery stores, and he was so good at his job that he was later promoted to run the chain's largest store, which was located in Winter Haven, Florida. George worked for Piggly Wiggly until 1930. By then, the real estate boom in Florida was a bust, but being an entrepreneur, George realized two very important things, first, that he was good at what he was doing, and second, that even in a depression people needed to eat.

In September of 1930, George ventured out on his own, and started his very own grocery store, which he called Publix Food Store, in Winter Haven, Florida. Things went so well that he opened a second location in Winter Haven in 1935, and eventually in 1940 he decided to open his dream store, which he called Publix Super Market. For 1940, this store stood out as something very different from other grocery stores of the time. The store has air conditioning, unheard of in other Florida grocery stores, and was equipped with fluorescent lighting, electric eye automatic doors, piped in music, in store donut and flower shops and other innovations that have since become the norm in grocery stores. Some referred to it as a "food palace." From that beginning, George proceeded to purchase 19 stores from the Lakeland Grocery Company in 1945 and began a systematic replacement program where he replaced those stores with larger supermarkets.

We have been a bit presumptuous in calling Mr. Jenkins, "George." All of his employees referred to him as "Mr. George" and he was well loved by his employees. Mr. George had a philosophy, treat your employees and your customers like family. That philosophy has served Publix well.

Today, Publix is one of the largest regional grocery chains in the United States, with 1,077 supermarkets and over $30 billion in annual sales. In the early 1950s, Publix moved its headquarters from Winter Haven to Lakeland, Florida, and today has stores as far north as Virginia, as far west as Alabama, and as far south as Key West.

According to Market Force's Customer Loyalty Index for 2017, Publix was voted America's favorite grocery store (tied with Wegmans), and has been in the top two of the Index for the past five years. Publix was named to Fortune's 100 Best Companies to Work For again in 2017, and has been on that list every year since Fortune created the list. Market Force Information, which complies that Customer Loyalty Index each year, said that in shoppers' minds Publix was the clear leader in the categories of "Ease of Finding Items" and "Availability of Items." In addition, Publix had the cleanest stores and fastest checkout speeds. In the category of Cashier Courtesy, Publix was second only to Trader Joe's, and second only to Wegmans in "Specialty Department Service." The only category where Publix was not at or near the top was cost, it is not the lowest cost grocery store, but it is obvious that cost is not the only consideration of a Publix shopper.

Keep in mind that the Index considers all grocery chains, including but not limited to Whole Foods, Winn-Dixie, Kroger, ShopRite, Hy-Vee, Fry's, ALDI, Food Lion, H-E-B, Albertsons and others.

Publix is the largest privately held company in the State of Florida, and as stated previously, has been on Fortune magazine's list of 100 Best Companies to Work For since its inception. Based on its revenue in 2014 (which has now increased), it was ranked the thirteenth largest retailer in the United States and thirty-fifth in the world. There are presently about 187,500 employees of Publix.

Why is Publix number one? Why is it so successful? Why does it beat out every other supermarket chain in America? We believe that it is because every Publix employee is a capitalist. Every Publix employee has the opportunity to be a stockholder of the company. Publix is the largest employee owned company in the world. Stock in Publix is only available to active employees and non-employee members of its Board of Directors. The stock cannot be sold outside the company without first being offered to the company for repurchase. Employees own about 80% of the stock, with the Jenkins family controlling about 20%.

The stock was first made available to employees in 1959, at $2.50 per share. The compounded annual growth from 1959 to 2016 was 16%, and the value of the stock as of March 1, 2017 was $40.90 per share. All of the great things about Publix are a result of the employees having a piece of the action, as Publix grows, expands and becomes more profitable, the employees share in its success.

With the need to more food products constantly to maintain freshness, and with the volumes of food that grocery store chains deal with, most people would assume that automation in grocery distribution centers would be very common, but that is an incorrect assumption. The vast majority of grocery distribution centers in America are old style systems, where people move products, drive forklifts and literally to all of the heavy lifting. Publix, of course, is one of the few that is different. Publix and a few others invested in automated storage and retrieval systems, referred to as AR/RS machines.

In 2015, Publix opened a new 1,000,000 square foot distribution center in Orlando, Florida. Publix hired a company called Swisslog to design their systems, which include stacker cranes, automated delayering machines and for its freezer facility, a two level split tray picking order fulfillment system. Publix is one of the companies that is showing how automation can be combined with human employees to assist them in their tasks, not to decrease the number of employees, but to increase productivity.

Publix lead the way in innovation in 1940, is continuing its tradition of innovation and is poised to lead the way with new innovations in robotics and artificial intelligence in the future.

Another George

Mr. George created an incredible supermarket chain in Publix, but there is also another George with an interesting story.

Although he is a fictional character, almost everyone has heard of George Costanza. This George was the loveable and hapless sidekick of Jerry Seinfeld on the hit television series "Seinfeld." In a famous episode of Seinfeld called "The Dinner Party" George shows off his new coat, which as made of Gore-Tex®. George talks about how cold it is outside and how warm his jacket is because it is made of Gore-Tex®. The story goes on and George gets punched to see if he can feel it through his Gore-Tex® jacket. As is typical with Seinfeld, there are repeated references to how warm the jacket keeps George.

So what exactly is Gore-Tex®? It is defined as a waterproof, windproof and breathable synthetic fabric. It is made out of layers of nylon, polyurethane, and PTFE (polythtraflouroethylene). PTFE is a fluoropolymer. Gore-Tex® is unique because it lets perspiration out, but does not let water in. Gore-Tex® was developed by, and is owned by, W.L. Gore & Associates, Inc., an American multinational corporation based in Newark, Delaware.

W.L. Gore & Associates, Inc. was founded by Wilbert (Bill) Lee Gore and Genevieve Walton Gore in the basement of their house in 1958. Bill had worked for DuPont for 16 years and had experience with fluoropolymers when he decided to start his own business.

Bill Gore's first project was to insulate a series of parallel electrical wires by using PTFE. His son, Bob, suggested a method which resulted in the company's first patent for Multi-Tet cable, a multi-conductor ribbon cable, which was used in computer and communications systems. Along with Gore-Tex®, some of the company's better known products are Glide® dental floss and Elixir® guitar strings. Using PTFE, Gore has created a variety of products for different industries, including medical implants, fabric laminates, filtration, membrane, venting, and sealant products and holds over 2,000 patents worldwide.

W. L. Gore & Associates also has a unique approach to business. None of its "employees" are called employees, they are "Associates." The company does not operate using a typical hierarchy of top down management, but is organized in teams. There are no managers and no one has a job title. The company uses what the founder referred to as the Lattice Organization, where there no chains of command, but a flat, lattice-like organizational structure where everyone was an "associate" and instead of bosses, there were Leaders, who associates would chose to follow. Compensation for associates is based on a peer level rating system.

The organization has consistently been listed on Fortune's 100 Best Companies to Work For, and W. L. Gore has now grown to over 10,000 Associates, working in 50 different facilities in the United States, Europe, Asia and Australia, with revenues of over $3 billion annually.

Like Publix, W. L. Gore is another employee owned company, with the Associates said to own approximately 90% of the outstanding company stock. Its success can be directly attributed to the pride, commitment and devotion of its associate-owners.

Jeff's Journey

Jeff was a relatively young man when he decided to wander around Belgium on a bicycle in 1988. Being quite an adventurer and also a beer lover, he rode through European villages testing different beers and dreaming of his own concoctions. He went home to Fort Collins, Colorado with visions of unique recipes that he could brew in his basement.

One of his first beers he named after the mountain bike he rode through Belgium, the one with the fat tires. Fat Tire was born, an amber beer that has been his company's trademark brew for many years. Of course, New Belgium Brewing Company would be no where without the love and attention of Jeff's wife Kim. While Jeff was brewing, Kim was doing everything else, bottling, marketing, distributing, financial planning, and general operations.

The basement brewing business became so successful that it was launched as a real commercial enterprise in 1992. The company has a purpose statement and a core values and beliefs statement which incorporates concepts like continuous, innovative qualify and efficiency improvements, promoting beer culture and the responsible enjoyment of beer, as well as trusting each other and kindling social, environmental and cultural change as a business role model.

The first new employee after Jeff and Kim became a shareholder in New Belgium, and the concept took hold and now stock ownership is open to all employees, which at last count was 770. All employees become eligible to be stockholders after one year of employment, and to add to the fun, each employee gets their very own cruiser bike on their one year anniversary. It's a different kind of place to work.

By December of 2012, the company was 100% employee owned. New Belgium says that their employees, as owners, look for ways to be less wasteful, more efficient, and to recycle and reuse. Once a year they have a strategic planning retreat, where input comes from all employees to help make decisions about the upcoming year's business plan. Through the company's policy of open book management, the company's finances are available to all employees, which they believe encourages a "community of trust."

Hardware Anyone?

While its exciting to talk about large companies with successful employee ownership plans like Publix, there are more than enough success stories about little companies that are generally unknown.

Jackson's Hardware has been hailed as a small company employee owned success story many times. The National Center for Employee Ownership lists Jackson's Hardware on its list of Great Employee-Owned Workplaces, and notes that Jackson's has also made the list of Winning Workplaces.

Jackson's plan was created in 1989, and stock was accumulated by employees through the plan for a number of years until it became 100% employee owned. The look and feel of Jackson's is an old-fashioned hardware store. Service is key, there is no hunting for someone to help you. An employee owner greets you, and makes sure that if they don't have an answer to your question, they go and get it from another employee. Since there are no cashiers, the same employee owner handles your purchase. Their newspaper advertising focuses on its 70 employee owners, and highlights their long years of service, typically including pictures of the group.

On the financial side, Jackson's is transparent. Sales figures are posted for employee owners to see. They all meet frequently to discuss ways to improve profits. Competitors have opened nearby but have had no real effect upon Jackson's. The level of service and attention has not been able to be met by the competition. It is rare for Jackson's to lose an employee, with most having served over ten years or more.

What does any of this have to do with robots and artificial intelligence? Hang in there and you will see.

CHAPTER VIII

TWO MEN, ONE PLAN

If your actions inspire others to dream more, learn more, do more and become more, you are a leader.

John Quincy Adams

Way back in 1956, two partners had a dilemma. They were both in their 80s and were looking to retire. The problem causing the dilemma was what to do about their company. They could sell it, but it was their baby, and they had seen similar companies in their industry being sold to larger companies, and disappearing into the black hole of a corporate giant. What they really wanted was to sell the company to its employees, because they knew that the employees had taken them where they were, would be good stewards of the future of the company, and loved it as much as they did and did not want to see the major changes that would come with a sale to an outsider. But how could that happen, when they knew the employees did not have the money for the purchase?

The partners hired accountants and tried to see if there was a way that the employees could afford the purchase. The answer was that even if the employees mortgaged their homes, there would not be enough capital to make it work. There must be another way, they thought.

Fortunately, they came upon Louis O. Kelso, a lawyer in San Francisco, and asked him to help them find a way to make this work. Kelso determined that the company had a profit-sharing plan that was IRS qualified and that the company had been making annual contributions into the plan. Kelso came up with a plan to use most of the profit sharing money to make a down payment of about 30% of the purchase price, and to have the profit sharing plan borrow the balance from a bank. The profit sharing plan would then continue to receive annual contributions and use those contributions to repay the bank debt. The contributions were tax deductible, and therefore both the principal and interest would be repaid with pre-tax dollars. The alternative was to have the company borrow the money and repay the loan with after-tax dollars. They could not afford the after-tax dollar plan, but the projections confirmed that the pre-tax repayment plan would fly.

Kelso then tackled the next problem. The IRS rules said that a related party transaction like this was a "prohibited transaction." Kelso did not let this stop him, since he found out that it might be possible for him to obtain an exemption from the IRS, which could be granted on a case-by-case basis if he could show the IRS that the plan was beneficial to the plan participants and was deemed to be "arms-length." Kelso successfully obtained the exemption from the IRS, and Peninsula Newspapers, Inc. of Palo Alto became wholly owned by its employees.

Employee ownership was nothing new in 1956. In 1921, stock bonus plans were permitted by the IRS to help motivate and incentivize employees. Large companies of the day like J.C. Penny, Pillsbury and Sears Roebuck instituted such plans, but until Kelso came along stock plans were not used as a methodology to create an exit plan for company owners, and the use of such plans by smaller companies was unheard of.

In 1958 Kelso and Mortimer Adler co-authored a book called "The Capitalist Manifesto." The book's focus was on employee ownership of businesses and set forth the argument for expanded use of tax approved plans to accomplish this goal. The authors argued that most employees had no way to acquire capital for investment and that the resulting wealth disparity was a negative force in society. Their most compelling argument is one that resonates with today's reality, that with technological advances, capital will become more productive and labor will be even more disadvantaged, which results in an increase in inequality. Kelso's arguments would seem to be even more pertinent in today's world than in the 1950s and 60s, since the technological advances of today make those years look like the stone age. In 1960 Kelso and Adler published their follow up book, "The New Capitalists."

According to David A. Spitzley, in *Democratic Capitalism Made Simple,* Kelso's Principles of Economic Justice can be summarized by the following three rules:

1. The Principle of Distribution: Each participant in the production of wealth should receive a share proportionate to the market value(s) of the labor and capital they contribute to the enterprise.
2. The Principle of Participation: Each household must have the opportunity to earn a decent standard of living through effective participation in the production of wealth, whether by property in labor, capital, or both.
3. The Principle of Limitation: No one may exclude others from effective participation in the production of wealth through excessive concentration of ownership, whether in capital, labor, or both.

As the concept of Kelso's plan spread, the IRS adjusted some of their procedures to accommodate these types of plans, and Determination Letters approving them became somewhat easier to obtain. The process hit a wall, however, in 1974 when Congress was working on the Employee Retirement Security Act of 1974 (ERISA) because early versions of this legislation did not provide for any mechanism to obtain an approval for such a plan. Critics believe that this was initially because Congress was completely unaware of this trend and unfamiliar with this type of employee stock plan. The law of unintended consequences was at work, and it appeared that this trend would end before it had even gotten fully started.

1974 was one of those watershed years. The economy in the United States was in the dumps, sitting at its lowest level since 1929. Tax rates were the highest they had ever been, interest rates were high and unemployment was over 10%. The economy needed a shot in the arm. Fortunately, Senator Russell Long, the chairman of the Senate Finance Committee, heard about Louis O. Kelso, and after meeting Kelso, decided that these so called "Kelso plans" were just what the doctor ordered for a sick economy. Senator Long championed changes to be proposed law which solidified the tax advantaged aspects of these plans.

As Kelso and others continued to use these plans, it became clear that the flexibility that could be built into a plan made them even more valuable. For example, if there are owners of different ages, it is not necessary to buy out everyone at the same time. An older owner who wants to retire could sell his or her shares to the plan and the remaining original owners could stay in place. Many owners also set up plans and then sell their shares over a period of years, without using any bank loans at all, just by making their annual contributions to the plan and having the plan purchase a certain percentage of the stock each year.

Based on what is happening in the world today, the name Louis O. Kelso may once again become a household word.

What Kelso dubbed the "second income plan" and others called the "Kelso plan" was primarily a concept designed to assist owners of businesses in selling their company to their employees. Kelso also advocated the use of such plans to finance corporate growth through the sale of newly issued stock to employees.

As these plans developed and adapted, one of the most vocal supporters of the corporate finance approach to the use of these plans emerged and added his voice to Kelso's.

Enter Walter P. Reuther.

Today few people recognize the name Walter P. Reuther, but throughout the 1940s to the 1960s he was one of the best known labor leaders in America.

Walter Reuther would seem to be an unlikely champion and hero of capitalism, since he spent many years of his adult life as the President of one of the most powerful labor unions in history. Reuther was born and raised in West Virginia, the son of a labor unionist and a socialist. He dropped out of high school at 16 and learned the tool and die trade. He later finished his high school education and attended Detroit City College. Reuther worked for the Ford Motor Company in the 1920s. Although early in his life he supported the socialist party, was raised by an avowed socialist, and went on an extended visit to the Soviet Union with his brother, later in life he became a fierce anti-Communist.

Walter secured various leadership positions in his local chapter of the United Automobile Workers union, and eventually in 1946 became the President of the UAW, and remained in that position until his untimely death in an airplane accident in 1970. He also became the President of the Congress of Industrial Organizations (CIO), and engineered the re-merger of the CIO with the American Federation of labor (AFL) in 1955.

Reuther's focus in life was social equality and justice. In the 1960s he was an ardent civil right activist and referred to by some as "the white Martin Luther King."

He was an advocate of the working man and believed firmly in the idea that employees ought to have an ownership opportunity in their workplace.

While Kelso was an advocate of employee ownership and believed that employees should participate in the benefits of being a stockholder, his focus to get them there was primarily through acquiring stock in a buyout of the owners. Reuther's focus was a shift in that thinking.

What if, instead of buying stock that had already been issued to an owner, the employee plan bought new issue stock directly from the company? The plan would borrow the money to buy the stock, and the money it borrowed would be used to expand and grow the company, instead of going to the former owner. This form of corporate finance would be the best of all worlds, the company could deduct the full amount of both the principal and interest of the loan, the employees would own some stock, and the company would get working capital. The concept of the employee plan as a corporate finance tool was born.

Not Coconut Grove, but Bohemian Grove

Bohemian Grove is a 2,700-acre redwood luxury "campground" located in Monte Rio, California, and it belongs to a private San Francisco-based club known as the Bohemian Club. The Bohemian Club's all-male membership and guest list includes artists, musicians, prominent business leaders, and government officials. It has been said that every Republican President since 1923 has attended their activities, along with various Democratic Presidents, and many cabinet officials. The industries represented by these people include military contractors, oil companies, banks (including the Federal Reserve), utilities, and national media. Some familiar names who have either been members or guests are Henry Kissinger, Thomas Watson Jr. (IBM), William Casey (CIA), George Bush, Walter Cronkite, and William F. Buckley.

In mid-July each year, Bohemian Grove hosts a two-week, three-weekend retreat for its members and guests. This retreat has sometimes been called an "informational clearing house for the elite." During the retreat, along with networking, there are speeches, known as "Lakeside Talks", where information and ideas are shared with the group.

In July of 1974, then California Governor Ronald Reagan addressed the Bohemian Grove Encampment, in a talk he called "Expanded Capital Ownership: The Only Answer to Creeping Socialism." The following is an excerpt from that address:

"...In our resistance to what some of us see as a creeping socialism, we have just theorized about the superiority of capitalism. Have we really made capitalism work to prove those benefits can do everything for everybody better than the promises of the socialists? All they can offer with their system, if you analyze it, is to take from the 'have' and give to the 'have-nots'. That doesn't eliminate have-nots – it just changes them around.

But capitalism can work to make everybody a "have". Some years ago, a top Ford official was showing the late Walter Reuther through the very automated plant in Cleveland, Ohio, and he said to them jokingly, "Walter, you'll have a hard time collecting union dues from these machines," and Walter said, "You are going to have more trouble trying to sell them automobiles." Both of them let it stop right there. But there was a logical answer to that. The logical answer was that the owners of the machines could buy automobiles, and if you increase the number of owners, you increase the number of consumers.

Over one hundred years ago, Abraham Lincoln signed the Homestead Acts. There was a wide distribution of land and they didn't confiscate anyone's already privately-owned land. They did not take from those who owned to give to others who did not own. It set the pattern for the American capitalistic system. We need an Industrial Homestead Act. . . I know that plans have been suggested in the past that all had one flaw. They were based on making the present owners give up some of their ownership to the non-owners. Now this isn't true of the ideas that are being talked about today.

Very simply, these business leaders have come to the realization that it is time to formulate a plan to accelerate economic growth and production at the same time we broaden the ownership of productive capital. The American dream has always been to have a piece of the action.

Income, you know, results from only two things. It can result from capital or it can result from labor. If the worker begins getting his income from both sources at once, he has a real stake in increasing production and increasing output. One such plan is based on financing future expansion in such a way as to create stock ownership for employees. It does not reduce the holdings of the present owners, nor does it require the employees to divert their own savings into stock purchases.

This one plan, and undoubtedly there are alternatives, utilizes an Employee Stock Ownership Trust to purchase newly issued stock when a corporation needs new capital for expansion. The trust acquires its funds by borrowing with a guarantee from the corporation, from a commercial bank, or other lending institution. Over a ten-year period, it is possible for 500 billion of newly-formed capital to be owned by individuals and families who today have little or no hope of acquiring a vested interested in our capitalist system . . . What a better answer could we have to socialism? What an export item on the World market! What argument could a foreign land have against a corporation which made its "have-not" citizens into "haves"?

In short, I am suggesting that we face a choice between government that has grown desperate, embarking on a course that leads to confiscation and redistribution, or using the great talent and expertise of the private sector to spread legitimate capital participation in free enterprise to those who now are only property-less employees…."

Ronald Reagan recognized that the answer to increasing dedication and opportunity for employees was not a give away, but an but a plan that showed them that if they invested their heart and not just their time, that they could earn the benefits of ownership along with a salary.

Walter Reuther testified before the Joint Economic Committee of Congress in 1967. At the time he was concerned about the loss of manufacturing jobs in America, and the fact that Japan seemed to be beating

American companies in head to head competition. His solution? Profit sharing for employees in the form of employee stock ownership was his conclusion. Before Mr. Reuther could see implementation of this approach, he died in an airplane crash, and labor and America lost its employee shareholders' champion.

Governor Reagan picked up where Reuther left off, and supported the concept of employee ownership. While the concept grew to where we now have millions of participants, it still remains one of the most underutilized corporate finance tools, and certainly the most beneficial, from the viewpoint of employee benefits and tax advantage.

CHAPTER IX

THE NUTS AND BOLTS OF EMPLOYEE STOCK OWNERSHIP PLANS

Bring on those tired, labor-plagued, competition-weary companies and ESOP will breathe new life into them. They will find ESOP better than Geritol. It will revitalize what is wrong with capitalism. It will increase productivity. It will improve labor relations. It will promote economic justice. It will save the economic system. It will make our form of government and our concept of freedom prevail over those who don't agree with us.

Russell B. Long

The predictions by economists Frey and Osborne that almost 50% of all occupations are threatened by AI automation technologies may just be another example of the overblown predictions discussed earlier, or they may be accurate. Knowing what we know about the resilience of the American entrepreneur, we believe that solutions will be found to use the new technology to the advantage of the overall society.

The President's Report, referenced previously, contains an excerpt from a speech by White House Council of Economic Advisers Chair Jason Furman, in New York, July 7, 2016:

"Fears of mass job displacement as a result of automation and AI, among other motivations, have led some to propose deep changes to the structure of government assistance. One of the more common proposals has been to replace some or all of the current social safety net with a universal basic income (UBI): providing a regular, unconditional cash grant to every man, woman, and child in the United States, instead of, say, Temporary Assistance to Needy Families (TANF), the Supplemental Nutrition Assistance Program (SNAP), or Medicaid. While the exact contours of various UBI proposals differ, the idea has been put forward from the right by Charles Murray (2006), the left by Andy Stern and Lee Kravitz (2016), and has been a staple of some technologists' policy vision for the future (Rhodes, Krisiloff, and Altman 2016). The different proposals have different motivations, including real and perceived deficiencies in the current social safety net, the belief in a simpler and more efficient system, and also the premise that we need to change our policies to deal with the changes that will be unleashed by AI and automation more broadly. The issue is not that automation will render the vast majority of the population unemployable. Instead, it is that workers will either lack the skills or the ability to successfully match with the good, high paying jobs created by automation. While a market economy will do much of the work to match workers with new job opportunities, it does not always do so successfully, as we have seen in the past half-century. We should not advance a policy that is premised on giving up on the possibility of workers' remaining employed. Instead, our goal should be first and foremost to foster the skills, training, job search assistance, and other labor market institutions to make sure people can get into jobs, which would much more directly address the employment issues raised by AI than would UBI."

We agree with Mr. Furman that the creation of a UBI is a step in the wrong direction, and will tend to create more problems than it would solve.

We believe that one of the ways to embrace the technology and simultaneously benefit workers is to turn them into capitalists who own a stake in the new automated companies. An underutilized tool that was created many years ago may be the perfect vehicle for this concept.

The Kelso plan, or the second income plan, has come to be known as the Employee Stock Ownership Plan (ESOP). The Employee Stock Ownership Plan is nothing new, but could provide a resurgence in popularity as the right tool to incorporate workers into the new technologies. What exactly is an ESOP?

An ESOP is simply a defined contribution employee benefit plan. The avowed purpose of an ESOP is to involve employees in the ownership aspects of the company. The ESOP is a very different concept from most defined contribution employee benefit plans, such as 401k plans, because the ESOP is required by law to invest primarily in the stock of the employer. As we all know, a 401k generally prohibits the purchase of the employers' stock.

What really makes an ESOP unique is that, unlike 401k plans and other benefits, it can be used as a corporate finance strategy. This is where it gets interesting, and where it could be employed as a way to help a company expand and incorporate AI automation, which simultaneously getting employees involved in ownership.

While ESOPs are most commonly used to provide an exit strategy for departing owners of successful privately-held companies, they can also be used to obtain financing for expansion and acquisition of equipment, like AI automation. ESOPs have the ability to take advantage of incentives to borrow money for acquiring new assets, such as AI automation, in pretax dollars.

When it borrows money, the ESOP becomes a "leveraged ESOPs" which can be used to not only reward and motivate employees, but reduce the sponsoring company's taxable income by the interest and principal borrowed to buy shares owned by the ESOP.

To begin with, a company that wants to start an ESOP needs to create a trust and appoint a trustee, which is typically the owner of other executive of the employer. Shares of stock of the employer company are then sold to the trust, for the benefit of the employee participants. Employees that meet the minimum requirements set for in the ESOP plan document can participate in ownership, which typically means full time employees over the age of 21 years. The stock is allocated within the trust to each participant employee's account. As is done in many stock option plans, when shares are added to employee accounts, they may not "vest" immediately. Under the law, employees must be 100% vested within three to six years. The time frame involved in vesting depends upon whether the shares are set to vest all at one time, or over a period of time. The details of this are set forth in each individual ESOP plan.

While we have discussed the fact that the trust is buying stock in the employer company for the benefit of the employees, we have not described how that happens, or where the money comes from to buy the shares.

In one scenario, after an independent appraisal is made of the value of the stock, the ESOP Trustee would obtain an agreement from the company to buy newly issued shares on behalf of the ESOP. The employer company would then borrow money from a lender, in what is sometimes called the "outside loan." The company would then immediately lend the loan proceeds to the ESOP in what is known as an "inside loan" so that the ESOP could purchase the stock.

How does the ESOP repay the "inside loan"? The company is required to make tax-deductible contributions to the ESOP (like making contributions to a profit sharing plan). The ESOP Trustee can then use the contributions to repay the inside loan, and in addition, the company may pay tax deductible dividends on the stock, which dividends can also be used to repay the inside loan.

The shares that have been purchased by the ESOP are held in a suspense account while the loan is outstanding, and as the loan is paid down, shares are released from suspense and allocated to the employee participants accounts.

The money paid for the shares to the company is used to purchase new AI automation equipment or for other expansion purposes. The end result of all of this is that the company, in essence, pays back the loan will all of the principal and interest on the loan being deductible. Why more companies are not doing this right now is beyond our comprehension.

There are different ways that dividends on ESOP stock can be deducted by the company. We just discussed the first way, which is to apply the dividends to loan payments in a leveraged ESOP transaction. If they are not applied to the loan payments, or the loan has been paid off, dividends may be paid in cash to ESOP employee participants, either directly or as payments to the ESOP that are distributed to participants within 90 days after the close of the plan year, or dividends many be voluntarily reinvested in company stock in the ESOP by employees.

When dividends are directly paid to plan participants on the stock allocated to their ESOP accounts, while such dividends are fully taxable, they are exempt from income tax withholding and are not subject to the 10% excise tax that applies to early distributions.

What companies should consider an ESOP? Although it can work for smaller companies, a safe place would be a value of at least $5 million, and a net income of at least $500,000 would be best. The company should have around 50 employees, but it could work with as little as 30, and the ESOP should consider purchasing at least 30% of the company. Does your company have steady cash flow? Is it increasing in value? Is it growing? Is expansion or AI automation viable?

ESOPs were created with the goal in mind that ownership opportunities and retirement assets would be available to employees of successful companies, and also to provide a method for owners of privately held companies to have tax incentives to sell their companies to employees. Most ESOPs have been created to provide a method for owners to sell out over a period of years to their employees who will carry the company into the future. The scenario above is purely an expansion plan using an ESOP, but it could be a combination of both an owner buyout and an expansion.

CHAPTER X
THE FUTURE

Looking ahead, future generations may learn their social skills from robots in the first place. The cute yellow Keepon robot from Carnegie Mellon University has shown the ability to facilitate social interactions with autistic children. Morphy at the University of Washington happily teaches gestures to children by demonstration.

Daniel H. Wilson

There is no doubt that the future is already here, and many people just don't realize it. The ability of machines to do many of the things mentioned earlier in this book is now possible, and will simply get more and more widespread.

So, what does the future hold? Let's take a look at the present and the past for some clues.

In some industries, things will happen very quickly but in many industries there is a resistance to change which will delay implementation of new technology, as we have described in the grocery industry. Another example that seems to make little sense is the lack of acceptance of technological improvements in the construction industry. Housing components and complete houses can be created faster, cheaper, stronger, more efficiently and with greater qualify control when built or partially built in a factory. Prewired wall units, preinstalled plumbing, and numerous other components can be created, yet the typical house is still built one piece at a time. Contractors and municipal building departments are accustomed to doing things one way, and have little or no interest in changing their view of how things are to be done. Another example of this attitude exists in the legal industry. A recent survey of law firms found that 65% of law firm leaders indicate that their partners are resistant to additional changes, yet 72% of those same leaders say that the pace of change will continue to increase. This is an issue that is difficult to overcome, and many of the predictions about changes in the workplace do not take into consideration the hesitancy of people to accept a new way of doing things. Many feel that "if it isn't broke, don't fix it."

Yet the Pace of Change Marches On

According to CB Insights, over 550 startups using artificial intelligence as a core part of their products raised $5 billion in funding in 2016, which does not include hardware focused robotics investments. Robotics startups had a record year in 2016, with 174 companies obtaining funding, and robotics investments alone between 2012 and 2016 accounted for $3 billion.

In the artificial intelligence world the following categories obtained venture capital investment in 2016:

Computer Vision / Image Recognition (Applications)
Computer Vision / Image Recognition (Platforms)
Context Aware Computing
Deep Learning / Machine Learning (Applications)
Deep Learning / Machine Learning (Platforms)
Gesture Control
Natural Language Processing
Personalized Recommendation Engines
Smart Robots
Speech Recognition
Speech to Speech Translation
Video Automatic Content Recognition
Virtual Assistants

As for robotics, there is also quite a variety of different focuses. In the Enterprise category, there were investments in the subcategories of drones (non-delivery), retail and warehouse, restaurant, heavy industry and manufacturing, laboratory automation, and delivery services. In the Consumer Category, there were investments in the subcategories of education, social robots for companionship, personal drones and consumer services like housekeeping. In the Medical category, there were investments in the subcategories of surgical robots, bionics and rehabilitation, and non-surgical medical tasks. In the Safety and Government category there were investments in the subcategories of building security, field assistance for law enforcement and the military, and drones, both aerial and underwater. There were a number of other robotics investments that do not fit neatly into any particular category.

What About Manufacturing?

It has been widely reported that since the year 2000 over 5 million American manufacturing jobs have been lost to a combination of moving those jobs overseas and automation.

The manufacturing process, particularly in the high tech industry, is a process involving multiple parties which all produce different parts. For example, only about 10% of the parts that go into an iPhone are manufactured in America. The rest are manufactured around a "hub" just like the situation was when America was funding massive amounts on space exploration. In those days, Cape Canaveral (later Cape Kennedy) was the hub, and hundreds of small to medium sized manufacturers grew up around the space center. Innovation frequently came from all of those contractors and subcontractors. Today, those "manufacturing hubs" are located in Japan, Taiwan, and China, among others, and while all of the suppliers to these hubs are not all necessarily clustered next to the major manufacturing plants, the concept remains the same. The closer the suppliers, the lower the cost of transportation, and just as importantly, the speed of acquisition of the supplies. Time in many instances is just as important as production cost. America now lacks those hubs, as well as those suppliers, and needs a plan to reinstitute them.

ESOPs are a well established, but highly underutilized tool, particularly for corporate finance.

In today's world, many people lament the fact that some of the most successful companies and the highest income generating companies operate with less jobs than successful companies of the past. The beauty of an ESOP is that it can be made to work for small companies as well as large ones. Specialized manufacturing companies may not need large numbers of employees, but we as a nation need a large number of specialized manufacturing companies. With automation, including 3-D printing, this is not only a possibility, but an inevitability.

Part of what America needs is a combination of tax reform to incentivize investment in manufacturing infrastructure and promotion of ESOPs to act as a corporate finance tool and additional incentive for companies to expand, while simultaneously energizing the employee base as shareholder-owners.

Tax reform as part of the answer

Industry sources report that more than $2 trillion in corporate earnings are parked overseas. The reason for this is that the United States has a tax rate of 35%, one of the highest corporate tax rates in the world. To counter this situation, many companies have resorted to tax inversions. Tax inversions are simply when an American company moves its headquarters abroad, avoiding United States tax while typically keeping executives stateside. Some pundits have even said that this the amount of earnings stashed overseas could be as high as $15 trillion.

For the companies that do not move offshore, the tax result is frequently that there is not sufficient capital to hire more employees or expand their business. Instead of companies moving offshore, it would benefit America to keep them here, tax them less, and have them create jobs for Americans, who, of course, pay income tax on their salaries. Lower taxes mean more capital to invest in equipment, such as robotics and artificial intelligence. With not just tax reduction in the form of lower tax rates, but tax credits for those companies that invest in their infrastructure, job growth and economic growth could be achieved.

We have learned that robotics and artificial intelligence cannot replace people, but what they can do is enhance the productivity of people. Most technology will not be designed to replace workers, but to replace some of the tasks that those workers are now burdened with, and will free them to do more "human" work, not just repetitive mindless work. If workers are freed to interact more with their co-workers in creative activities, leaving redundant work to the technology, productivity, along with job satisfaction will increase. In addition, with workers owning part of the business, statistics have consistently shown that productivity increases, employee turnover decreases, and wages typically increase.

Education

As we alluded to in Chapter IV, it has been estimated that 65% of children presently entering primary schools will end up working in new job types that do not exist today. How do we prepare them?

From the information available, the trend appears to be going toward machines that perform tasks in a human type manner, but we must always remember that as human as they may seem, these machines cannot replace us.

We have discussed the fact that manual labor and low paying jobs may be the first to be replaced by automation, but is that really what will happen? We make the assumption that to recreate what we believe to be simple tasks should be relatively easy for advanced artificial intelligence. This appears to be untrue.

A researcher in the 1980s named Hans Moravec introduced what is now called Moravec's Paradox. This paradox is simply that while a computer may perform tasks like playing checkers, it cannot do simple things like bending down and picking something off the floor, or other basic sensorimotor skills. These skills require a substantial amount of computational resources, even more than figuring out which checker to move to which square. While we think all of those jobs will go away, we are probably wrong on some level. Some of our movements and actions are done without much contemplation, and are natural for a human, but very difficult for a robot. What we are seeing is that artificial intelligence is very good at a narrow area of work, so the goal is to get the machines to do what they are good at, and let the humans do the rest.

Common Sense

It seems to me that common sense is not so common anymore. Many people seem to lack what we always referred to as common sense. An example of this is the lawsuit filed by a woman in California against a jelly bean company because they did not disclose to her that jelly beans contained sugar. Really? We could all cite hundreds of other examples.

My wife believes that the schools should now teach a course in what she calls "practical intelligence," which a polite way of saying common sense without insulting the people you are teaching. That phrase, in and of itself, exemplifies what only humans can do, ie: give something a nicer name to spare the feelings of the people to whom it is directed.

Based on experience, the future belongs to the ones who will have a combination of skills. Along with the old "reading, writing and arithmetic" the new workers will need reading, writing, arithmetic and computer science. Those will be the basic underlying skills that all workers will be required to master. In addition, on top of those basics, and to have a chance at being successful, future employees will need social skills, interpersonal communication skills and practical intelligence.

In the past, we would all see mothers and fathers constantly interacting with their children. Today we see mothers and fathers with their faces buried in the smart phones, texting or talking or reading. The children are doing the same as their parents. Children now text from their bedrooms with questions like "what's for dinner?" Many young people have no manners and no social skills, and when they throw a temper tantrum at their job, they won't understand why they were fired. Of course, this is a generalization and not all young people act this way, but certainly a substantial percentage fit into this category. I suspect that schools teaching practical intelligence will become commonplace.

Human Attributes

We believe that various human attributes will be difficult if not impossible to translate to machines. For example, creativity is a uniquely human characteristic. The ability to formulate ideas is human. How do we stimulate creativity in young people, which will be an attribute that will be in great demand in the future? In addition, creativity needs to be centered in a framework of values, human values.

We believe that teaching young people values will be the most important task to guide future generations and will help them succeed in becoming employed and productive.

In schools today, it does not appear that any form of values are being taught. In the public schools in particular, the attitude appears to be "it's not my job."

Values will complement artificial intelligence since AI has no human values of its own. For western civilization to survive and thrive alongside artificial intelligence, the human side of the equation must follow a consistent pattern of common sense moral values.

What we see as human attributes are based on a series of human values. The attributes that humans have, like creativity, empathy, motivation, and nurturing, are derived from common sense values.

In America and most of the western world, our traditional values are based on Judeo-Christian ethics. Those ethics include things like the dignity of human life, honesty, a strong work ethic, common decency, respect and consideration for others, gratitude and generosity,

How ESOPs fit into this plan

According to the Center for Economic and Social Justice, less than a dozen ESOPs were in existence in 1965. Today over 10,000 companies have adopted some form of an ESOP and there are somewhere between 11 million and 13 million employee shareholders in America. In about 1,500 companies, the majority ownership is controlled by the employees.

While Walter Reuther was a huge advocate of ESOPs, much of organized labor has been skeptical about employee ownership, but some labor unions have strongly endorsed the idea. Many groups in labor unions do not believe in mixing what they call "labor and management." We think this is a shortsighted approach, since we believe it to be beneficial for labor to have a say in management, and as owners they would have that right and opportunity.

An ESOP has many facets. Primarily it acts as an incentive for employees to become and remain more productive, since they have the pride of ownership and share in the upside of the company. It is also part of a retirement program, as the employee builds equity in the company over years. An ESOP typically buys back an employee's shares upon retirement out of plan funds. It acts as a tax deferred method for employees to build wealth.

In some companies, additional plans are also created, such as 401(k) plans that go hand and hand with the ESOP. Companies that provide both an ESOP and a 401k provide diversification of investment, since an ESOP buys stock in the employer and a 401k invests in outside companies.

An ESOP can provide cash and stock bonuses to employees. A company can deduct dividends paid on ESOP stock, with the dividends either going to repay an ESOP loan or be paid out directly to employees if there is no loan, or after a loan is retired. Some plans are arranged so that any dividends got into purchasing more stock for the employees to increase their share of ownership, which benefits them upon retirement.

On the corporate finance side, an ESOP is unique. There is no other financial vehicle in America that has its benefits. Only an ESOP can protect individual shareholder-employees from personal risk in the event of default because ESOP loans are not guaranteed by the employees. Only an ESOP can borrow money for corporate expansion or growth and pay back the entire amount of both the principal and interest in pre-tax corporate funds. This is absolutely unheard of under any other scenario, and gives the ESOP a massive advantage over any other method of financing.

The National Center for Employee Ownership (NCEO) has identified what it calls the "ownership edge." While ESOP ownership creates an advantage, that advantage is enhanced and bolstered by a specific ownership culture that successful companies have implemented. According to the NCEO companies with the ownership edge comply with six essential rules. As stated by NCEO, those rules are as follows:

1. Provide a financially meaningful ownership stake, enough to be an important part of employee financial security.

2. Provide ownership education that teaches people how the company makes money and their role in making that happen.

3. Share performance data about how the company is doing overall and how each work group contributes to that.

4. Train people in business literacy so they understand the numbers the company shares.

5. Share profits through bonuses, profit sharing or other tools.

6. Build employee involvement not just by allowing employees to contribute ideas and information but making that part of their everyday work organization through teams, feedback opportunities, devolution of authority, and other structures.

This approach is essentially different from the typical management approach, and some company owners may not be comfortable with this management style. However, if company management wants to survive and thrive in the new economy, serious consideration of this approach needs to be taken.

The support for ESOPs over the years has sometimes occurred in a surprising context. The story of the Polish trade union Solidarity even has a connection.

In August of 1980, shipyard workers in Gdansk, Poland formed Solidarity in an effort to defy the Communist regime and obtain fair wages for the workers. At first, the government tolerated Solidarity, but as it grew stronger, the government's patience wore thin. Eventually, Russia had enough and ordered its puppet government in Poland to put a stop to Solidarity. Martial law was instituted, and the leaders of Solidarity, including its founder Lech Walesa, were arrested.

President Reagan and Soviet Premier Leonid Brezhnev traded threats, and Reagan instituted sanctions against Poland, although our allies in NATO declined to participate. What happened next was more than interesting. In what some later called the "Holy Alliance" Reagan joined forces with Pope John Paul II, the first Polish pope, to pressure Poland and Russia. The Pope told Russia that he would fly to Poland and stand with Solidarity. The Pope and the President had a plan, get as much equipment into Poland as they could to help spread the word of freedom and empowerment to the workers, in the form of computers, radio transmitters, printing presses, fax machines, copy machines and other communications devices. The CIA and priests worked hand in hand to smuggle equipment into Poland. Without a shot being fired, Lech Walesa was eventually released and the puppet government fell.

In September, 1990 President Reagan travelled to Gdansk, Poland to celebrate the tenth anniversary of the founding of Solidarity. In a speech outside the shipyard were Solidarity was created, he spoke these words:

"…Meanwhile, what about the workers in those state monopolies that are being put up for sale? I am reminded of a technique for employee ownership that has worked well for many U.S. companies. It goes by various names but the best known is "Employee Stock Ownership Program" or "E.S.O.P.". With such a program, the employees of a company create a trust which borrows money from a bank to buy shares of stock in the company. The loan is paid back over several years from the employees' share of the company's profits. How can they be sure the company will be profitable? The workers, as owners, make sure by insisting that unprofitable or obsolete products be replaced by new ones; that operating costs be kept down; and that new efficiencies of operation are adopted. In the U.S. we have seen it happen time and again.

It is another fact of human nature: When a person owns assets – a house, land, a small business or shares of stock in a big one – he or she will look after those assets…"

Reagan supported the ESOP concept, for America and for others. He saw it as a technique to help those formerly communist countries establish a basis for capitalism by getting workers to be owners and to share in the growth and profitability of their companies by becoming capitalist like Americans. He also believed that more American companies should take advantage of the unique attributes of an ESOP.

While it is encouraging that we have gone from a very small number of ESOPs in 1965 to a situation where between 11 million and 13 million Americans participate in ESOPs, we can do better. America will be stronger and better if 50 to 75 million employees, or more, are part owners of their employer. We need to strive for more employee shareholder participation.

We all know that changes are coming and automation is increasing, but instead of fearing the future and predicting massive job losses, we should remember the words of Ray Kurzweil who said "You can point to jobs that are going to go away from automation, but don't worry, we are going to invent new jobs. People say "What new jobs?" I don't know. They haven't been invented yet."

Kurzweil almost sounded like he was quoting Doc Brown, the character from the *Back to the Future* movies when he said "the future has not been invented yet."

Remember that humans (particularly Americans), unlike machines, are resourceful, creative, opportunistic, and self-serving. I suspect they will find something to do, and it will be something profitable.

EPILOGUE

Can ESOPs solve all of our economic problems? Of course not, but it seems logical that the expanded use of the ESOP concept would be beneficial for most employees as well as their companies. As well has this concept works, it is a surprise that it has not been better utilized.

Today the statistics show clearly that the middle class family has lost a substantial percentage of its purchasing power over the past few decades, which translates into less savings for retirement. Simultaneously, many of the larger corporations have experienced large percentage increases in executive compensation, frequently with no corresponding increase in the value of the stock of their companies.

We are not advocating the dreaded "universal basic income" which we believe sends the message that some in our society have just given up, and provides further incentive for a defeatist attitude for others. Instead, we believe that the incentive of some ownership position in an employee's company may be what is needed to inject new life into those of us in the 99%.

Presently less than 10% of American workers are employed in businesses that offer the ability to become an owner through an ESOP. We believe that the economic impact of increased ESOP ownership to the level of 20 or

25% would jump start economic activity and make true capitalists out of many employees. All of this results in a more secure future for employees and more productivity for America.

As technology has spread throughout the world, and in particular, America, production has increased to amazing levels. As we alluded to earlier in this book, over 90% of Americans were involved in agricultural activities 150 years ago, and only 2% are so involved now. The amazing part about that statistic is that America is producing so much food with the 2%, that we are exporting enough to literally feed many other countries. While not as obvious, similar changes have occurred in other industries. Technology is not something to be feared, but something to put to work for the benefit of mankind.

Our hope is that with AI and robotics, production will continue to increase and profits will increase, providing more benefits to the entire population, and perhaps particularly to employee owners of the companies that embrace the change.

OTHER BOOKS BY

ROBERT C. HACKNEY

AVAILABLE ON AMAZON.COM

Let's Bring America Back
Lawyer's Guide to Blockchain Technology
Entrepreneur's Guide to the
New Equity Crowdfunding Rules
Mergers & Acquisitions 101

To contact the author, please write to
fourpalmspublishing@gmail.com